William

(315) 638

7/19/2000

The Cambridge Quintet

Also by John L. Casti

Paradigms Lost
Searching for Certainty
Complexification
Five Golden Rules
Would-Be Worlds
Boundaries and Barriers: On the Limits to Scientific Knowledge,
edited with Anders Karlqvist (Addison-Wesley)
Science as Storytelling:
The Creation of Science in Myths and Stories,
edited with Anders Karlqvist (forthcoming, Addison-Wesley)

The Cambridge Quintet

A Work of
Scientific Speculation

John L. Casti

Helix Books

Addison-Wesley
Reading, Massachusetts

Library of Congress Cataloging-in-Publication Data
Casti, J. L.
    The Cambridge quintet : a work of scientific speculation/John L. Casti.
        p. cm.
    Includes bibliographical references.
    ISBN 0-201-32828-3 (alk. paper)
    1. Artificial intelligence—Philosophy.   2. Philosophy of mind.
    I. Title.
Q335.C375   1998
006.3'01—DC21   98-11757
                                                              CIP

Addison-Wesley is an imprint of Addison Wesley Longman, Inc.

Published in Great Britain by Little, Brown and Company

Jacket design by Suzanne Heiser
Typeset in New Baskerville by the author

23456789-DOH-0201009998
Second printing, March 1998

Find Helix Books on the World Wide Web at
http://www.aw.com/gb/

To the memory of Alan Turing and John von Neumann,
creators of the modern computer-age

Contents

The Cambridge Quintet

Author's Note

The book you hold in your hand is not a novel; but it is a work of fiction, part of an emerging genre that I like to label 'scientific fiction.' The Japanese term for this kind of work is a *shōsetsu*, a rather more flexible and generous term than 'novel.' Such a work, while containing elements of fiction, is more of a chronicle; in this case, a work attempting to convey in a fictional setting the intellectual and cognitive issues confronting human beings involved in shaping the science and technology of their future. Had the book been a piece of conventional popular science writing, I would have been limited to what is known about the motives and thoughts of the people involved, while if a work of science fiction or a general novel had been my goal, then the story would have had to adhere to the principles and conventions of those genres, concentrating on the development and change of world views of the book's characters through the resolution of conflicts. But in scientific fiction the overriding goal is very different. It is to present a lively and comprehensible exposition of the intellectual and emotional uncertainties involved in shaping the future of human knowledge. So in this sense, scientific fiction has as its mission to try to imagine how the world we live in today was shaped by decisions of the past, and how the decisions we take today impinge upon the world of the future.

The conflict explored here is a conflict of ideas, pitting Ludwig Wittgenstein and Alan Turing on opposite sides of the issue: Can a machine think? A fictional dinner party seems a good venue in which to speculate about how these two titans might have argued the matter, as well as to periodically inject the thoughts of the other thinkers at the party on a panoply of related topics bearing on the nature of human cognition and the possibility of mechanical thought. To paraphrase the well-known 'mediaologist' Marshall McLuhan, in this book the ideas are the message.

The dawning of a new intellectual age is always a time of excitement and ferment. In these periods of transition from the old to the new, many competing ideas are being bandied about and even the most powerful intellects get caught up in the cross currents blowing across the field just coming into existence. The beginning of the discipline we now call 'artificial intelligence' is no exception. So it should not come as a surprise to the reader that the fictional account of the hypothetical—but possible—gathering presented here will on occasion see the participants making statements that in ways depart from what we might imagine they would have said on the basis of their published works.

It's a well-known fact of life, academic or otherwise, that people often say things at dinner parties that they would be loath to put down in writing the day after. This is normal. It's another fact of intellectual life that decades after the occurrence of a momentous event, especially after the participants in the event have been elevated to the status of icons, we see the event and the people from the perspective of what has happened during the intervening decades—or centuries—and not from the perspective of the time of the event itself. Such is the situation with the story told here. Readers familiar with the philosophical arguments and technical developments in the thinking-machine debate over the past fifty years will see the issues discussed here in a totally different way than how even intellectual giants such as Snow's guests saw them when the issues were fresh and untrammeled by the vague musings and personal prejudices of later philosophers, computer scientists, and neurophysiologists. Things look a lot different in the AI business today than they looked in the summer of 1949. This is a point to keep in mind when assessing the imagined views of the participants as I've presented them in the pages of speculation that follow.

A last caveat: For the sake of exposition, I have moved several conceptual themes in AI from their actual time in the post-1950 decades back to the period of this dinner. The reader should not infer from this that I am arguing that ideas such as Noam Chomsky's theory of language acquisition or John Searle's notorious Chinese Room argument were really developed by any of the dinner guests. It is purely a speculative issue of *imagining* that these ideas had been put forth at that time. How might the dinner-party participants have reacted? The book's final section corrects all such achronologies, and gives pointers to further reading on these and other matters discussed in the body of this narrative.

Finally, a word of thanks to the many people who helped in the preparation of this book. For their advice on the overall idea, as well as comments on the initial proposal, I would like to express my gratitude to Greg Chaitin, Kirk Jensen, George Johnson, Jeff Johnson, Melanie Mitchell, Tor Nørretranders, and Jeff Robbins, as well as to the book's original editor, Eamon Dolan. Readings of the penultimate version of the manuscript by Doyne Farmer, Atlee Jackson, David Lane and John Wyver saved me from many a *faux pas* in both language and content. To each of them, a hearty thanks for an unenviable job well done. Finally, kudos of the first magnitude to the book's editor, Richard Beswick, who always helped and never hindered, supporting me in those inevitable dark moments during the course of any book, when it looks as if the project will never be completed.

Dramatis Personæ

C. P. Snow (1905–80) Novelist, civil servant and physicist, Snow illustrated with his own success that his proclaimed division of Western society into 'two polar cultures'—the sciences and the humanities—need not be absolute. Snow earned a doctorate in physics at Cambridge (1930), where he was a Fellow of Christ's College. He recruited scientific talent for the Ministry of Labor during World War II, and later served as a member of Parliament and of the Cabinet. His Rede Lecture at Cambridge in 1959, "The Two Cultures and the Scientific Revolution", warned of the consequences of the lack of communication between scientists and humanists. Snow was made a life peer in 1964.

Alan Turing (1912–54) Mathematician who in 1936, while an undergraduate at Cambridge, published an article in which he created a theoretical machine that could move from one state to another by following a prescribed set of rules. This 'Turing machine' led to a computing scheme that foreshadowed the logical structure of modern digital computers. During World War II, Turing played a leading role in efforts to break enemy codes. He later worked on the development of the first electronic computers, on theories of artificial intelligence, and on the applications of mathematics to biological forms. Turing was arrested in 1952 for violation of British homosexuality statutes, and committed suicide at the age of 41. The 1987 play *Breaking the Code* by Hugh Whitemore is based on Turing's life.

J. B. S. Haldane (1892–1964) Geneticist, science popularizer and political activist, who helped to bridge the gap between classical genetics and evolutionary theory with his mathematical analyses in the field of population genetics. After studying at Oxford, Haldane spent ten years at Cambridge before taking a chair at University College, London in 1933. In addition to his purely scientific work, Haldane was a devoted Marxist and for many years headed the editorial board of *The Daily Worker*, the newspaper of the British Communist Party. Haldane became disillusioned with Communism following the Lysenko affair in 1948. He emigrated to India in 1957, where he continued his work on statistics and genetics until his death.

Erwin Schrödinger (1887–1961) Nobel prize-winning physicist, famous for his work in quantum mechanics. Following a doctorate at the University of Vienna in 1910, Schrödinger succeeded Max Planck in the chair of theoretical physics in Berlin in 1927. He emigrated from Germany in 1933—the same year he shared the Nobel prize for physics with Paul Dirac—because of Nazi threats. In 1939 Schrödinger joined the newly formed Institute for Advanced Studies in Dublin, where in 1944 he set the stage for what is now molecular biology in his lecture series "What is Life?". Schrödinger spent his later years exploring a life-long interest in the foundations of physics and their implications for philosophy and Eastern religious thought.

Ludwig Wittgenstein (1889–1951) Perhaps the most influential philosopher of this century, who was unique in the annals of philosophy, having developed two completely different philosophies during his lifetime, the second of which entirely repudiates the first. Wittgenstein began to study the philosophy of mathematics with Bertrand Russell at Cambridge in 1912, work which led to his masterpiece the *Tractatus Logico-Philosophicus,* written during his service with the Austrian Army in World War I. Having given away a large inheritance, Wittgenstein taught elementary school in Austria in the 1920s, returning to Cambridge to resume his philosophical work only in 1929. In 1939 he was appointed to the chair of philosophy formerly held by G. E. Moore, resigning this post in 1947 to devote his final years to writing up his many ideas. Wittgenstein's work on language, the foundations of mathematics, logic and meaning shed much new light on a variety of problems, notably skepticism and the problem of other minds.

Prologue

The Story Begins

The revolution began in the early afternoon on a lazy English summer day in 1935 when Alan Turing, an undergraduate at King's College, Cambridge, had the idea for a theoretical gadget by which to settle the Decision Problem, a famous outstanding question in mathematical logic. At about the same time, heated debate in the Commons Room of the Princeton mathematics department over another mathematical tangle led to the development of a new kind of logical calculus, one that put the heuristic notion of what it means to carry out a computation on a sound mathematical footing. A decade later, motivated by his code-breaking work during the Second World War, Turing, along with John von Neumann and others in England and the United States, began the process of translating these abstract mathematical notions about computation and logic into actual computing devices.

By the mid 1940s, the practical, day-to-day benefits of computers were clear for all to see. But the scientists leading the way to the development of computers, especially Turing in England and von Neumann in the United States, were already beginning to reflect on the ultimate capabilities of such machines, including their potential to carry out many of the tasks that had hitherto been considered the private preserve of human beings. The existence of these 'computing machines' resurrected a plethora of classical psychological, philosophical, sociological and linguistic puzzles about the nature of human nature that are still as fresh and timely as on that fateful day when Turing invented his 'Turing machine'. And primacy of position on this list of conundrums is the eternal question: 'What's so special about human beings?' One way to sharpen this question is to ask: 'Can a computing machine be as good as a human in its cognitive capacity?' Even more generally: 'Could a machine ever be developed to the extent that we would accord it full human rights?'

The difficulties in coming to terms with such an essentially philosophical question reside as much in clarifying what we mean by 'thinking', by a 'machine', and by the word 'human' as it does in any particular notion of intelligent behavior. In this sense, the issue of thinking machines belongs squarely in the realm of philosophy—but with a twist. The feature separating the problem of thinking machines from other philosophical puzzles like, 'What is truth?' or 'What is just?' is that one can at least imagine constructing a physical device whose behavior is cognitively indistinguishable from that of a normal human being. Or so thought Alan Turing, anyway, upon completing his code-breaking duties in Bletchley Park at the end of World War II.

To separate the fact from the fantasy in these sorts of speculations, here we hypothesize that in the summer of 1949 Sir Ben Lockspeiser, Government Chief

Scientist, and Sir Henry Tizard, science adviser to the Ministry of Defense, discussed the question of thinking machines with the celebrated novelist and physicist (and later government spokesman for science and technology) C. P. Snow, asking him to sound out the scientific community as to the likelihood of this 'transspeciation' coming to pass. Snow's response was to arrange an informal dinner at Christ's College, Cambridge, his alma mater, to which he invited Turing, along with several other seminal thinkers whose diverse backgrounds and interests all bore upon the general question as to whether machines could ever be built that would actually think. What follows is a speculative account of the ideas circulating around the dinner table that June evening in 1949.

Chapter One

The Sherry

An Evening at Christ's

The tall, balding, avuncular man in the slightly rumpled suit and horn-rimmed glasses looked like nothing so much as a droopy-eyed basset hound as he bustled about his old rooms at Christ's College, instructing Simmons, the manservant, as to exactly where to place the tray of glasses and the bottles of sherry, whisky and water and, in general, reliving a bit of his life here as a student. Yes, Charles Percy Snow was in his element again—at least for tonight. Simmons had seen it all before, of course, and managed to bear up under Snow's mixture of fidgety nervousness and nostalgia with the stoicism typical of the British serving class. It was good indeed, he thought, to have Mr Snow back in college again, if only briefly. Just like old times. Pity he seemed so preoccupied with this dinner tonight. It must be some very important people he's expecting, mused the servant, as he set the drinks and glasses on the sideboard.

As he went about supervising arrangements for the dinner, Snow thought back to a recent discussion with Sir Henry[1] about this chap Turing and his group at Manchester. According to Sir Henry, they were in the process of trying to build a machine that would eventually be capable of really thinking, just like a human being. While sharing Sir Henry's skepticism about the possibility of a machine ever doing anything remotely like writing a novel such as *War and Peace,* or even solving a simple problem in logic, Snow agreed that the potential implications were so enormous that the government should look into it if there was the slightest chance that it might be carried off. Rather clever of Sir Henry to suggest that I arrange this dinner by way of getting to the heart of the matter, thought Snow. All the academic specialties and scientific and philosophical expertise gathered around the table tonight should certainly be able to shed a bit of light on whether Turing's vision of a thinking machine are just academic fantasies or have some basis in reality.

Bleak, dreary, miserable postwar Britain, grumbled John Burdon Sanderson Haldane, his mood growing blacker by the moment as it kept pace with the unseasonably cold and rainy late-spring Cambridge weather. As he entered Christ's College from St Andrew's Street, Haldane glanced up at the two intricately carved yales gracing the college's turreted gateway. They seemed to be gazing mournfully down at him, their sad, antelope-like eyes conveying the impression that they, like Haldane, grieved for an empire that would never be again. Or perhaps, reflected Haldane, they were only empathizing with a fellow being caught out in a Cambridgeshire

---

[1] Sir Henry Tizard, chief science adviser to the Ministry of Defense.

thunderstorm when by rights he should have been tucking in to a piping-hot steak-and-kidney pie and a tot of whisky at his local pub. Pondering that pleasant vision, but fleetingly, JBS passed through the gate into the college's First Court on his way to more serious matters.

Moving across the First Court, Haldane's thoughts shifted quickly back to the rather more immediate concern of what his old friend Percy Snow could possibly have had in mind when he insisted that Haldane catch the next train up from London for 'a bite with a few friends' at Snow's old rooms at Christ's. If it was so damnably important, why couldn't they have met down in London? Percy's tight-lipped refusal to disclose his reasons for the dinner—beyond saying that it was something 'just up your street—a lot of science and a little philosophy'—was irritating, to say the least. And the miserable ride up from London didn't make it any better. Never much of an admirer of British Railways, even when it ran on peacetime schedules, Haldane wondered more than occasionally whether the Ministry of Transport knew that the war was over. One would never guess it by the service tonight, he fumed. By 1949, at least the trains should be back on schedule, he believed, even if the rest of the country wasn't. Indeed, it would be fair to say that JBS's mood was at least as foul as the weather and showed about the same likelihood of lightening up anytime soon.

As he hurried past the Master's Lodge on his way to the Fellows' Building at the back, the tall, burly, bald-headed Haldane gave the appearance of a playfully alert but rather ill-tempered walrus, an impression acquaintances said was only accentuated by his bushy, sandy-colored mustache, brusque manner and barking tone of speech. And his spiky temperament contributed to some of Haldane's detractors referring to him as 'that prickly cactus' during corridor conversations in his laboratory at the University of London.

Approaching the Fellows' Building, Haldane again wondered about Snow's tantalizing remark: 'a lot of science and a little philosophy'. Since when did government mandarins like Snow give a fig for philosophy? And since when did His Majesty's science advisers start asking geneticists for advice on anything—*especially* philosophy? Damned odd, thought Haldane, as he pushed through the heavy oaken door and started up the staircase to Percy's rooms.

The pain's electric fingers probed his abdomen like living things, seeming to reach for the essence of his very soul while blotting out the dull, grey Cambridge skies and the bustle of dons, students and tradesmen on Sidney Street. Ludwig Wittgenstein paused for a moment to lean against the corner of a building, as he tried to push the pain back into a small corner of his mind where it could be controlled, if not conquered. As he caught his breath, he recalled the sad events of the past few weeks, feeling again the burden of the terminal cancer that had claimed his eldest sister Hermine in Vienna, the very same scourge that now seemed to have a death grip on his own life as well. His normally intense, penetrating gaze and even complexion had given way to the haunted, hollow-cheeked look and pale coloration of a medieval saint in an El Greco painting. The way the disease seemed to be progressing, he thought, it would require a miracle for him to complete the dictation of his thoughts on language games before leaving Cambridge later in the month.

And what was he to make of this puzzling dinner invitation from the novelist Snow, a man he had never met and whose writings he found long-winded, tiresome and far too 'British' to be taken seriously? Whatever had possessed him to accept this strange invitation? Perhaps it was Snow's remark that tonight's dinner may well open up a whole new chapter in modern philosophical thought. Typical literary bombast, thought Wittgenstein. But he had to admit to himself that it piqued his curiosity to think that his work on the philosophy of language and mind had come to the attention of even a minor man of letters like Snow.

But tonight's dinner might still prove interesting, he reflected, especially if Snow delivered on his promise that Turing would be there. He hadn't seen Turing since the younger man had attended his lectures on the philosophy of mathematics in the spring of 1939. And even though they had had their differences then about the nature of mathematical truth and what it meant to carry out a 'computation', Turing had certainly done some first-rate work since those days on the nature of computing machines and their relationship to epistemology and the mind. Wittgenstein smiled to himself, then, as the pain began to ease a bit, resumed his trek along Sidney Street on his way to Christ's.

Odd how the course of one's life is dictated by seemingly minor, even inconsequential twists and turns of fate, reflected the dapper, frizzy-haired gentleman in the grey tweed suit as the train to Cambridge finally pulled out of London's Liverpool Street station. Last month I spoke on the wireless for the BBC about free will, human thought and the indeterminism underlying quantum theory. Now I find myself going to meet one of my listeners to discuss a matter he describes as being 'of the greatest national importance'. Surely this fellow Snow must know that I'm a foreigner, and in no position to undertake any sort of secret work for His Majesty's government. It would have been difficult to refuse the invitation, though. Besides, a novelist and politician of Snow's stature should at least set a decent table and have invited some congenial companions for conversation, perhaps even an attractive lady or two, thought the man, always on the lookout for new challenges—and conquests—both intellectual and personal.

By 1949, the 'dapper gentleman', Professor Erwin Schrödinger, was one of the most celebrated and publicly visible physicists in the world. As a principal architect of the quantum theory of matter, he was the recipient of the 1933 Nobel prize for physics, and currently headed the Dublin Institute for Advanced Studies. Schrödinger had recently embarked on an entirely new path of scientific investigation, one that involved the study of the physical basis of living organisms. As the train wound its way through the outskirts of London, he recalled Snow's cryptic remark during their short chat on the telephone to the effect that these new-found biological interests were a key aspect of the matter he wanted to discuss tonight. Pity he hadn't pressed Snow a bit more on this point, as it might have shed light on what possible connection there could be between the physics of a living cell and Snow's mysterious issue 'of great national concern'. But no matter. It would all get sorted out in just a few hours' time, thought Schrödinger, as lost in thought he stared out at the flat expanses of the East Anglian countryside.

Good heavens, muttered the guard to himself, as the wiry, dark-haired man entered the carriage. There's something just plain undignified, if not downright suspicious, about a man wearing a pyjama top under a sports jacket—especially when the jacket doesn't look as if it's been cleaned or even pressed since it left the shop, observed the guard, as he moved down the carriage checking tickets.

It would no doubt have come as quite a shock had the guard known that the 'suspicious-looking' man in the rumpled sports jacket, nervously twisting and bending his second-class ticket to Cambridge, was one of the prime contributors to the Allies' recent victory over Germany, a man unknown to the general public but regarded as something of an eccentric genius in scientific circles.

Alan Turing had done his war service at Bletchley Park, a rural estate midway between Cambridge and Oxford, working as a code-breaker. When it became known, early on in the war, that the German military was sending coded orders to its forces using a machine termed the Enigma, a handful of mathematicians, led by Turing, analyzed methods of using intercepted messages and various searching techniques to pin down the workings of the Enigma machine. These scientists developed strategies that eventually led to their being able to decipher messages as if they were receiving the uncoded text directly from the German High Command. By the end of the war, Turing had had enough daily contact with electronic machinery and its uses for uncovering patterns in data to begin to think seriously about building a computing machine that could actually *duplicate*—if not exceed—the thought processes of the human mind. And it was this very notion that occupied his mind that afternoon on the circuitous routing the train had to take to get from Manchester down to Cambridge.

Oblivious to the guard's disapproving glances, Turing spent most of the trip staring up at the ceiling, reflecting on the Lister address delivered earlier in the month by his Manchester University colleague, the famed neurosurgeon Sir Geoffrey Jefferson. How bloody wrong-headed can a man be, fumed Turing, to think that because a machine isn't built from biological parts like flesh and bones, and doesn't have emotions like a whimpering dog or a laughing baby, it's not capable of rational thought? Old Jefferson has really put his foot in it this time, Turing giggled to himself, by setting out these emotionally laden and completely baseless arguments claiming that if a machine can't write a sonnet or compose a concerto, then it can't display intelligent, human-like behavior. The man actually seems to believe that thought comes from what the brain is made of, not from how it actually works. One might as well believe that a wrist watch made of steel and glass can't tell time because it doesn't have a swinging pendulum and a wooden case like a grandfather clock. How could the BBC even broadcast such a moronic argument?

Turing hoped that tonight's dinner would help set things back on course. Snow's promise to bring together some influential people for a sensible scientific discussion of the possibility of building an intelligent machine would be bound to help straighten things out, he felt. This man Snow certainly appeared to have all the right political connections. And his claim that the Ministry of Science and the War Ministry were both keen on the Automatic Computing Engine (ACE) was a most encouraging sign, thought Turing, as he made a mental note to himself to try to speak with Snow in private at some point during the evening about obtaining support from the government to build the ACE.

*   *   *

Turning away from the oak-paneled and beamed dining room of his suite of rooms, Snow looked through the doorway connecting the dining room to the drawing room, casting his glance up at the plaque over the fireplace in the Georgian-style drawing room. It commemorated another Charles, Charles Darwin, who had occupied these very same rooms during his tenure as a Fellow of Christ's more than a century and a half earlier. How great is the privilege of those chosen to 'come up to Cambridge, to follow in the footsteps of illustrious men, pass through the same gateways, sleep where they had slept, wake where they had waked', thought Snow, recalling the timeless words of Wordsworth. And how fitting it seemed that tonight's conversation would take place in rooms that had housed the man who almost single-handedly catapulted the study of humankind from the subjective, emotional realm of theology into the objective, rational domain of science.

Snow felt that the discussion this evening would almost certainly turn upon the question of what makes a man a man and not a machine, and that was certainly a theme that Darwin would have entered into with considerable enthusiasm. However impervious one might be to the feeling of past time, Snow believed that there were moments when one was drugged by it. It is a kind of haze that envelops one in these rooms, he mused, as one glances out at the college chapel, touches the ancient oaken panels, or gazes out over the roofs to King's. If Darwin were dropped into the First Court today, thought Snow, he would feel instantly at home; everything has been the same here for so long that Snow wondered if it would ever change—and hoped that it wouldn't.

Sinking down into one of the window seats in the drawing room overlooking the First Court, Snow allowed himself to slip into a moment's reverie, as he cast his mind back to the many evenings he had spent in this room listening to accounts from Allberry[2] of the pitfalls of translating Coptic script, or trying to converse with Trend,[3] his overly excitable neighbor down the hall, a man who seemed always capable of beginning his sentences in English but quite unable to complete them in anything other than Spanish or Portuguese. But most of all Snow thought about Hardy,[4] whose death just over a year ago had come as such a painful blow. In his mind, he could almost hear Hardy's soft voice saying once again, 'Mark that man,' as in his inimitable fashion he once again asked Snow to judge the cricketing skills of one of the new men at Fenner's.[5] Those early-summer outings to Fenner's with Hardy, followed by a few games of Stumpz[6] after the evening meal, were among Snow's most treasured memories of his time in Cambridge. How simple and somehow *purer* life had seemed in those halcyon days before the world went up in flames.

Hello! What's that racket in the hallway? Snapping out of his reverie, Snow jumped up from his armchair and moved to the doorway, where the pounding and stomping about had by now reached epic proportions.

"Well, Haldane. From the racket in the passage and

---

[2]C. R. C. Allberry, an Orientalist and Fellow of Christ's, who was killed serving in the RAF during the War.

[3]John Brande Trend, Professor of Spanish and Fellow of Christ's College.

[4]Godfrey Harold Hardy, famed mathematician, Professor of Mathematics, Fellow of Trinity College, and one of Snow's closest friends at Cambridge.

[5]The Cambridge University cricket ground.

[6]A cricket board game.

the drumming on the door, I should have guessed it would be you. Come in, man, and get out of that wet coat. How was the trip up from London?"

"Beastly, if you must know," snorted Haldane. "Overcrowded cars and late departures don't say much for British Railway's prospects of getting back on to a peacetime schedule. Can't say it helps my disposition much either," he added sourly.

Casting a quick glance around the drawing room before removing his mackinaw and black homburg, Haldane moved to one of the window seats and looked around the suite of rooms, as if searching for a sympathetic ear upon which to focus his heartfelt grievances against the weather, British Railways—or both.

"It appears I'm the first to arrive. Must say I found your invitation a bit odd, Snow. What is all this business about computing machines, minds and philosophy? How do these things have any bearing on 'national interests'?"

Letting this tirade roll off his back in silent amusement, Snow wondered what kind of bee had found its way into JBS's bonnet tonight. To settle his friend down, Snow smiled enigmatically and said, "All in good time, all in good time. Let's wait until the other guests arrive before going into these matters. Meanwhile, may I offer you a drop of sherry?"

"Just the ticket, my friend. And unless I'm mistaken, it looks as if you have a pretty decent bottle of Amontillado there on the sideboard. It pays to be a Fellow of a rich college like Christ's, eh, Snow?"

"Honorary Fellow now, actually," Snow answered. "But the Master sometimes humors us old boys."

"Unless I'm mistaken, more of your guests have just arrived," said Haldane, as Snow handed him a glass of sherry.

Putting his own glass down on the side table, Snow moved to the hallway. As the oddly disorienting cadence of words being spoken in a foreign language came floating in, Snow opened the door and found both Wittgenstein and Schrödinger arriving at the same moment. "Ah! Our Austrian contingent," he said. "Very good. I thought I heard the sound of German out in the hallway. And right on time, too. Nothing I admire more in a man than punctuality. Please come in."

Austrian nationality, expatriate ways and a fondness for matters philosophical are about the only things Schrödinger and Wittgenstein have in common. Schrödinger, the son of an industrial chemist and amateur botanist, cuts a rather dashing figure in a three-piece grey tweed suit, his light-blue eyes flashing behind the scholarly, wire-rimmed spectacles that have become his trademark in photos showing the group of physicists—Heisenberg, Bohr, Pauli, de Broglie, Born, Dirac and Schrödinger—who created the quantum revolution. These spectacles sit below a high forehead and light-brown frizzy hair, a small part of the surface persona of a man who exudes a strange combination of great personal charm and total self-preoccupation. This is perhaps not without a small touch of irony in someone noted for his complex and irregular sexual life, centered on the delicate balancing act of cohabiting with both his wife and his mistress in the strict Catholic environment of Dublin. A completely areligious man in the Western sense, Schrödinger has long rejected any kind of ethical system based on collective concerns, arguing both in person and in print for a personal variant of the Vedantic notion, which says the self and the world are one—and they are all there is.

By way of contrast, Wittgenstein has spent his entire life bound up in the moral struggle of trying to be what he terms a 'decent' man. For him, this has been an ongoing battle to triumph over the temptations of pride and vanity to be dishonest. Unlike Schrödinger, Wittgenstein is imbued with a fundamental religious morality, wrapped up in the belief that to understand ethics one must see the world as a whole. This conviction, in turn, had led to his service as a hospital orderly during the Second World War and to an expression of deeply felt sympathies for the concerns of the working class. As he shuffles painfully into the drawing room, Wittgenstein appears pale and somehow shrunken, projecting the look of a man resigned to a death that wouldn't be long in coming.

"Please help yourselves to drinks from the sideboard," offered Snow. "We have a rather decent sherry from the college cellar. Or perhaps you'd prefer something a bit stronger?" Schrödinger poured himself a generous measure of whisky, adding a splash of soda. Ignoring Snow's invitation to the bar, Wittgenstein, a man congenitally unable to abide the chit-chat and social banter common to such situations, drifted over to the other side of the room.

A bit taken aback by Wittgenstein's unexpected ill manners and oddly withdrawn behavior, Snow moved over next to him and they both stared out of the window for a few moments at the Master's Lodge across the courtyard, lost in their own thoughts. Uncomfortable with the prolonged silence, Snow finally expressed his condolences on the death of Wittgenstein's sister.

"Yes, these past months have been difficult for our entire family," replied Wittgenstein softly. "Cancer must surely be one of the most painful of deaths, and Hermine suffered greatly at the end. Thank God it's all over now," he said pensively, as he continued to stare out at the darkening early-summer evening and the rain beating down into the courtyard.

As Schrödinger joined them at the window, Snow congratulated him on his recent election as a Foreign Member of the Royal Society. "Quite an honor. But then, I don't suppose it's quite the same as the thrill of receiving the Nobel prize, is it?" Snow asked.

"It's always a source of satisfaction to be acknowledged by one's peers with any type of honor or prize. It's strangely discomforting for me, though, being placed in the position of a prophet without honor in one's own land," Schrödinger remarked.

"May I infer that you're considering returning to Austria?" enquired Snow.

"Not any time soon, I'm afraid. The war may be over, but the same dangerous fools are still running the country. And these Nazis are like elephants, my friend. They have long memories. They won't soon forget my speaking out against them during my brief stay in Graz—and neither will I."

Ignoring Schrödinger's bitter recollections of the recent Nazi period in Austria, Haldane stepped into the conversation and asked Wittgenstein about a rumor circulating to the effect that he had moved to Ireland since resigning as G. E. Moore's successor to the Chair of Philosophy in Cambridge two years earlier.

"I've been staying in Dublin since I left Cambridge," replied Wittgenstein, "although at the moment I'm visiting friends here in Cambridge."

"What are you working on now?" asked Haldane.

"Mostly I'm occupied with arranging my thoughts of the past few years for publication. I'll be off in just a few weeks to visit Malcolm, my former student, who is now a professor at Cornell University in the United

States, and I want to finish dictating my notes before I leave."

Pausing for a moment to reflect on the way in which his ideas on language had changed over the past twenty years, Wittgenstein mentally reviewed his position that language was a public or social phenomenon. He then explained that "In these notes I intend to stress that language can function only if the rules are accepted by more than one person. Language, you see, is a publicly available social reality, not some kind of essence whose nature can be worked out in your head by pure reasoning alone."

"I'm not a professional philosopher," said Haldane, "but isn't this view of language diametrically opposed to the one you advanced earlier in your book *Tractatus Logico-Philosophicus*? There you seemed to argue that the world must consist of simple objects that can relate to each other in certain ways—independently of humans and of language. This would seem to say that the world consists of facts, which are merely rearrangements of simple 'atomic' objects. Or am I mistaken?"

"Yes, at that time I did feel that all meaningful statements could be analyzed or decomposed into sets of statements that are like pictures of configurations of objects. But if this were indeed the case, then the analysis of such statements is the right method of philosophy. Meaning is then turned into a picturing relationship. But I now repudiate this view of meaning as a tool. I'm convinced that the meaning of a statement is simply the sum total of all the ways the statement can be used."

Wittgenstein then illustrated this new view of meaning. "Consider," he said, "what I call a 'language game'. There are two extremely important features of games that are relevant to the way language is used in everyday life. The first is that games are rule-governed practices, while the second is that they are related to each other by a sort of family resemblance. Chess and checkers are good examples of this."

"Well, what if I have a toothache?" asked Haldane. "How do you make a picture of that?"

"It's just plain wrong to think that when we say something like 'I have a pain', that pain is a definite, identifiable inner object that we notice within ourselves and report to other people. We can't talk of our mental life as if we are reporting private experiences. But that's what my earlier picture theory would lead one to believe. Now I say that behavior and circumstances are integral to the understanding of how we talk about mental life."

"If I've understood you correctly," said Haldane, "your conclusion is that it's not possible to talk about our knowledge of the mind in the old-fashioned Cartesian manner. We cannot blithely assume that the contents of the world are of two utterly different sorts: an outer world of solid, visible objects in space and time, and an inner world of thoughts and feelings. Your new theory of language would say that these two worlds overlap closely, and that there can be no speaking about inner thoughts and feelings that is not connected to the manifestation of thoughts and feelings in the circumstances in which they occur. Yes?"

Before Wittgenstein could reply, Snow stepped in to break up this learned discourse. "Well, Wittgenstein," he asked, "tell me, how are you finding life outside the academic womb of Cambridge?"

This was clearly a topic that Wittgenstein had considerable energy to pursue, as he quickly turned to face Snow, his eyes seeming to have at least momentarily regained their well-known sparkle and intensity as he animatedly snapped off a reply: "Infinitely preferable. Academic life is detestable. The gossip of my college bedmaker far surpassed the insincere cleverness of the High Table repartee. Einstein was right to say that to do real intellectual work one would be better off earning a livelihood by being a shoemaker during the day, and doing one's real thinking by night."

Despite being an eternal outsider to the British academic establishment himself, Haldane was taken aback at the intensity of Wittgenstein's outburst against the academic world, and tried bringing the discussion back on to less contentious ground, remarking to Wittgenstein that "Some of my Marxist colleagues in London mentioned that some time ago you were planning to move to Russia and take up a teaching job in Moscow. Any truth to that?"

"I did actually visit Russia in 1935. But I wanted to help them build Communism by working on a collective farm, not by teaching in the city. My concerns are with the working class, not the intelligentsia."

"What happened?" enquired Schrödinger.

"What happened was that I discovered life in Russia was much like being a private in the army. Nothing could destroy my sympathy with the Russian regime faster than the growth of the kind of class distinctions I saw emerging there. It's enough to make Britain look positively egalitarian by comparison. I really am a Communist at heart. The problem for me is that the Russian regime isn't."

"I've visited Russia myself a number of times," said Haldane, "and have been very favorably impressed by the treatment of scientists and science. But I can't say I really accept Stalin's brand of Communism; it's just too exclusively economic."

"Think of the spiritual purity that comes from the Communist ideal, Haldane. It's the working class on the farms and in the factories that will build a healthy, moral society, not the scientists in their labs or the bureaucrats and generals in their offices," intoned Wittgenstein.

With some asperity, Haldane replied that "Until this Lysenko business last year, I could not have disagreed with you more. But now I'm not so sure. Stalin's interference with science and his treatment of anyone who questions Lysenko's idiosyncratic theories of genetics is forcing me to re-examine my enthusiasm for Marxist ideology. I really can't stomach the jackboot manner in which Lysenko's opponents are being stomped upon."

"Tyranny doesn't make me feel indignant," said Wittgenstein. "What's important is that the people have work."

The man is a bundle of contradictions, thought Haldane, unable to bring himself to reply to such an outlandish statement. Turning away from the conversation, he moved over to the sideboard to refill his glass.

Meanwhile, Snow and Schrödinger had drifted over to the other side of the room and fallen into a deep discussion of the problem of free will that Schrödinger had brought up a few weeks earlier in his BBC broadcast, and during which he had addressed the role of electrons in human thought processes.

Schrödinger explained that in his view "The uncertainty arising from the quantum-mechanical nature of the electron has nothing whatsoever to do with the problem of free will versus determinism as it pertains to human behavior. And this despite the unquestioned fact that the source of all behavior ultimately rests with electrical activity in the brain. But I must say that I haven't been doing much real physics lately. Instead, my thoughts are coming back to my original interests in philosophy. One never forgets a first love, it seems."

"How true," agreed Snow. "But this work on the relationship between electrons and thought sounds as if you're doing as much biology as philosophy. I've seen your little book, *What is Life?*, and the issues you raise there about the physics of living systems would seem to be most germane to our discussions tonight."

"Which brings us back to the point, Snow. Just what *have* you got us up here for?"

"I told Haldane a moment ago that I'll tell you everything—in due course. I do think, however, that you'll find the trip to have been well worth your time. If nothing else, the college kitchen has promised us a decent meal. And I dare say that's not something to be taken lightly these days. There's one more guest yet to come, however, and I see by the clock that like most mathematicians he's a bit on the casual side when it comes to keeping appointments. But I'm sure he'll be along shortly."

Joining Snow and Schrödinger, Wittgenstein confronted Snow, stating rather pointedly, "Your invitation mentioned a matter of potentially great national importance. I can't think of a single thing of interest to me as a philosopher that could possibly be of concern to His Majesty's government. A philosopher is not a citizen of any community of ideas. That is what makes him a philosopher. And whatever it is you have in mind, you can count me out if it has anything to do with the military. Now that the war is over, I'm not willing to contribute a moment of my time to this bloody 'Cold War' building up between the Americans and the Russians. I hope that's not what you've brought us together for this evening."

"My dear fellow," replied Snow, "allow me to put your fears at rest on that score. Each of you brings a very special background and point of view to this gathering, one that is of the utmost importance to the issues I want to put before you tonight. While the military may well take an interest in our deliberations, there are no immediate—or even long-term—military implications to the matters I want to lay on the table this evening. This is purely an informal meeting of minds, arranged so as to get your views on some highly speculative, and I must say basically philosophical, issues."

At that moment a faint tapping was heard from the passageway. Haldane boomed out, "Let's hope that's your missing guest, Snow. I'll have a look."

Throwing open the door, Haldane found himself facing a slightly built man in a raincoat rather the worse for wear, beneath which the top of a pyjama jacket peeked out from under a sports jacket. Completing the picture of a quintessentially disheveled academic was a pair of flannels that looked to be precariously held up by nothing more than a piece of twine or, perhaps, string. "Ah, Dr Turing, I presume?" joked Haldane.

Like most mathematicians, Alan Turing was basically an introvert, far more at home with abstract symbols and the inevitability of chains of logical argument than with the unspoken ambiguities, innuendos and enervating vicissitudes of human affairs. As Haldane and Snow ushered him into the room, Turing stuttered an apology for his tardy arrival, mumbling half-heartedly something about problems with the train connections down from Manchester.

"Nothing like British Railways to get a man to his destination too late for drinks," sympathized Haldane.

"We were beginning to wonder about you, Turing," said Snow, somewhat peevishly. "Good of you to make it down here tonight. May I offer you a sherry? But wait! Perhaps you're in training for one of those long-distance foot races I've heard you compete in. If that's the case, perhaps you'd prefer something a little less alcoholic?"

"Yes, thank you. Some water or a glass of soda would be very nice," Turing replied in a quiet, almost timid, voice.

"Are you acquainted with the other guests—Schrödinger, Haldane, Wittgenstein?"

"I know Wittgenstein and Haldane, of course," said Turing, looking up rather shyly. "Actually, I heard Wittgenstein's lectures on the foundations of mathematics some years ago. But I've never met Professor Schrödinger," he said, turning to the man from Dublin.

"My pleasure, Turing," smiled Schrödinger, shaking the younger man's hand with some enthusiasm. "I've admired your mathematical work for quite some time now. I'm very pleased to finally have the opportunity to meet you."

Turning to greet Wittgenstein, Turing broke into his characteristic stutter. Somewhat embarrassed, Wittgenstein extended his thanks for a reprint of an article on the mathematics of computation that Turing had sent him, and mentioned that he had been following work in this area with considerable interest. He then asked Turing about his activities during the war.

"I spent most of it developing methods for breaking codes," Turing replied. "That activity started me thinking about how one might construct physical devices that could carry out the types of theoretical computations outlined in my earlier work. So now I'm preoccupied with the theoretical and engineering problems surrounding the construction of this kind of 'computing machine'."

"Do you mean that you are constructing an actual physical machine that follows rules so as to carry out computations?" Wittgenstein wondered aloud in a tone of faint disbelief.

"Yes, that's exactly what I'm doing. Just one year ago, one of my colleagues in Manchester wrote the first set of rules ever for such a machine. We call these rules a *program*," Turing stated proudly. "A later program succeeded in computing the largest factor of a very big number."

"How big exactly?" asked Haldane.

"I don't recall the particular number. It's of little consequence, anyway. That calculation was just to test the workings of the machine. What's truly important, though, is that we now have a machine that can carry out computations just like a human 'computer'. In my view there's even considerably more to it than that. Doing a numerical calculation like finding the largest factor of some number is simply a special case of a far more general kind of operation, the manipulation of symbols. In my opinion, this is the essential ingredient in human thought itself. So our hope is soon to be able to build a machine that can really think like a man."

"Complete rubbish," cried Wittgenstein. "It's nonsensical to apply the term 'thinking' to any kind of machine—whether it's a computing machine or a steam engine. Thinking requires mental states. And the property of having a mental state is completely tied up with the hustle and bustle of everyday human life. You're mixing up two entirely different things here, Turing, when you speak about carrying out mathematical calculations with a computing machine and when you talk of these machines actually thinking."

Sensing the start of a heated discussion, Snow came over to separate the combatants, thrusting his hands out in a placatory gesture that implored both Turing and Wittgenstein to hold their arguments for later. "Gentlemen, please," he pleaded, "I see you've already come around to one of the key issues I've asked you here to consider. So may I suggest we retire to the dining room and continue this discussion over dinner?"

"I could do with a bit of sustenance," chimed in Haldane. "What do you think, Schrödinger, shall we move to the dining room?"

"Suits me perfectly. But I want to hear more, much more, about Turing's 'thinking machine'. How is it built? What principles of physics and mathematics does it rely upon? And how does it actually *think*? I won't leave these rooms tonight until Turing gives us a full account of all these matters."

Chapter Two

The Soup

Brains and Machines

A cheery fire crackled in the fireplace beneath the plaque commemorating Darwin, taking the chill off the unseasonably cool evening. The guests situated themselves around an elegantly set rectangular oaken table laid for five: Turing sat across from Wittgenstein, while Schrödinger took the place opposite Haldane. Pulling out the chair at his own place at the head of the table, Snow instructed Simmons to bring on the soup, a rich, creamy lobster bisque.

"Fine-looking soup, Snow. Compliments to the college kitchen," proclaimed Haldane, whose spirits seemed to have brightened considerably since the appearance of the servant with the soup. Reaching for a slice of bread from the basket on the table, he noted that "It certainly looks as if wartime shortages are a thing of the past at Christ's College—in the kitchen, anyway."

"If only that were indeed the case," said Snow sadly, as he glanced owlishly over the top of his glasses at

Haldane. "Fortunately, though, as a courtesy to His Majesty's government the Master has been generous enough to dip into the college reserves for a few of the items on this evening's menu."

He then turned to address the table at large. "Since you all seem curious about why I've asked you up here this evening, let me finally come around to the point of our gathering."

Snow began to lay out the reason for the evening's dinner by summarizing some of Turing's wartime work on the German Enigma coding machine, along with earlier theoretical work by Turing in Britain and others, including von Neumann[1] and Church,[2] in America.

"This work has convinced some of the government's science advisers of the feasibility of constructing powerful computing machines," Snow observed. He then went on to note that "In fact, one of the first such machines is already in operation in Manchester. At present, these machines look as if they'll be exceedingly useful for solving certain types of problems arising in mathematics and the natural sciences—code-breaking, calculating the flow patterns of fluids, determining the trajectories of planets and the like. As interesting and important as all this work is, it isn't what I've called you up here to discuss tonight. Rather, I would like your frank views as to whether these machines are likely to be useful in more general cognitive tasks of the sort we usually associate with creative human thought."

---

[1]John von Neumann, Hungarian-born American mathematician. Developer of the theory of games, a good bit of mathematical economics and numerical weather forecasting, in addition to fundamental contributions to pure mathematics. Inventor of the notion of a stored program for the digital computer.

[2]Alonzo Church, American logician. Developer of the lambda calculus, a logical language that is mathematically equivalent to Turing's procedures for formalizing the notion of a computation.

Shifting the thrust of his opening remarks to the theoretical background underlying these computing machines, Snow then gave a brief account of the relationship between computation and mathematics. "In the late 1920s, it was taken for granted in mathematical circles that every well-posed mathematical question must have a definite answer—true or false. For example, suppose I claim that every even number is the sum of two prime numbers. Among the mathematical literati this is an assertion known as Goldbach's Conjecture. The conventional wisdom of just a few decades ago would have it that a well-defined mathematical proposition like this must necessarily be either true or false. Moreover, there should exist a chain of logical reasoning that would lead in a finite number of steps to which of these two possibilities is actually the case. So thought the mathematicians of that day, anyway."

"And so think most people still today," piped up Turing, "including the majority of scientists."

"Indeed," continued Snow. "But in 1931, the Austrian logician Kurt Gödel proved that the mathematicians were wrong. He showed that in every logical system with enough expressive power to allow us to state all possible propositions about numbers, there must exist at least one statement that can be neither proved nor disproved following the logical rules of that system."

Turing interrupted again. "Exactly. Gödel proved that not every mathematical question has to have a yes-or-no answer. Rather, a question, even a simple one about numbers, may be simply undecidable. In fact, Gödel proved even more. His work showed that there exist questions that while being undecidable by the rules of the logical system, can be seen to actually be true if we jump outside that system—they just cannot be *proved* to be true."

31

"Thank you very much for helping to clarify these points, Turing," remarked Snow drily.

Schrödinger then asked, "Isn't such a fact simply a translation into mathematical terms of the famous Liar's Paradox? Something like the statement, *'This sentence is false'*."

"Exactly," responded Turing again, seemingly not able to rein in his enthusiasm for discussing these mathematical questions.

"If you'll allow me to continue," said Snow, "I promise that we will soon return to a detailed discussion of these fascinating issues."

Turing finally leaned back in his chair and fell into a sulky silence, as Snow continued: "It's important to keep in mind that a crucial assumption on which Gödel's results rest is that the logical system we use must be consistent. This means that by using the logical operations of the system, it's not possible to establish that the same proposition is both true and false. This then raises the question of how we can know whether a logical system is or is not consistent. Gödel answered this question very succinctly too: we can't! He established the fact that no logical system can prove its own consistency."

Bowing his head in Turing's direction, Snow went on. "Shortly before the War, Mr Turing here found a way to translate these logical results about numbers and mathematics into analogous results about calculations and computing machines. Perhaps now I can give the floor over to Turing, who I'm sure will be pleased to give us all a brief account of his own work in this area. Turing?"

"Thank you very much," said Turing, who by then was almost beaming at the prospect of describing his results to such an illustrious group.

The Soup

"In 1935, I attended a course of lectures on mathematical logic given by Max Newman here in Cambridge. At one point in the lectures, Newman spoke about Hilbert's[3] Decision Problem. This problem asks if there's a single logical framework sufficient to prove or disprove *every* mathematical statement. Gödel's work of just four years earlier had destroyed forever Hilbert's belief that such a logical framework must exist. But I looked at the problem from a totally different point of view than Gödel. My idea was to consider the logical steps one goes through in constructing a proof as being the same as the steps a human calculator would follow in carrying out a computation."

"Some people," said Snow, taking a kind of perverse pleasure in being able to interrupt Turing for a change, "including Mr Turing, are convinced that this type of mathematical problem-solving is merely the tip of an iceberg, in the sense that the real significance of such machines is their ability to duplicate human thought. But this notion seems far-fetched to many."

"Yes," interrupted Haldane. "I heard Sir Geoffrey Jefferson's Lister address a few weeks ago on the BBC, in which he held that machines are incapable of creative thought because their components are not biological."

"Perhaps so," continued Snow. "That's one of the issues we'll want to examine here tonight. But if there's even the smallest possibility that computing machines may one day be developed that can actually think, that fact would be of such enormous significance for every aspect of human life that His Majesty's government feels the idea should be looked at seriously."

[3]David Hilbert, Professor of Mathematics at Göttigen University in Germany. One of the most famous mathematicians of the century. Leader of the school of mathematical philosophy termed "formalism", which was destroyed by Gödel's work.

33

Snow concluded, "So I've asked you all up here this evening to give me your frank and open views on this matter. Is the construction of a thinking machine possible, even in principle? Or are there logical, philosophical and/or technical obstacles that will prevent us from ever constructing such a device? Following our deliberations, I'll try to distill your collective wisdom into a report on the matter to the Minister of Science. But before we jump into this question of thinking machines, I think it may help put the issue into perspective if Turing would again take the floor and kindly bring us up to date on how his computers work, as well as on why he feels that these machines bear any resemblance at all to the human brain—in either its structure or its function. Turing?"

Setting his spoon down and placing his hands flat on the table, Turing looked down and began to speak in a soft, stammering, barely audible voice. "A mechanical computer," he said, "is basically a large number of address locations called a *store*, or a *memory*, together with an *executive unit* that carries out the various individual operations involved in performing a calculation. These operations are what we call a *program*. Such an operation, or step in the program, might consist of the addition or comparison of two quantities stored at different locations. Let me illustrate the idea using these empty soup bowls and the spoons here on the table."

Grabbing the bowls from in front of Schrödinger, Haldane and Wittgenstein, Turing set them in a row in front of his place and went on. "Let's imagine that I want to use the machine to add the numbers 1 and 2. The computing machine would begin with all the bowls being empty, as they are now. To add 1 and 2, I first place one spoon representing the number 1 into Haldane's bowl and two spoons for the number 2 into

Wittgenstein's. The machine then consults its program for addition, which would have an instruction saying, in effect, 'Take the spoons from Haldane's bowl and place them in Wittgenstein's bowl. When you finish this operation, take the resulting set of spoons in Wittgenstein's bowl and move them into Schrödinger's bowl. Finally, print the number of spoons in Schrödinger's bowl on to a tape.' This is the way a computing machine adds 1 and 2 to give 3. It's clear, I think, how the process mimics the steps a human calculator would go through to carry out the same sum."

"But surely your computing machine must have more to it than a few soup bowls and dirty spoons," interjected Haldane with an air of incredulity. "If this is all that's involved, it would be a damn sight easier to add 1 and 2 by hand on a piece of paper—or even in my head—than to call upon one of your machines."

"Indeed," said Turing. "But there is a lot more to it than this. The soup bowls and spoons are only to give you a feeling for how simple the basic operations of such a machine really are."

"Why don't you give us just a few more details?" Snow requested. "I think we'll need to understand a bit more about these machines and their operation if we're going to be able to see any connection between the way they work and the operation of human brains and thought processes."

Taking up Snow's suggestion, Turing went on with his account of how a computing machine functions. "One might picture such a machine as a kind of gigantic post office with many postal boxes. Taking this analogy one step further, the instructions to be executed by the computer can be thought of as being carried out by a postal clerk working under the supervision of the postmaster. In fact, a computing machine actually contains a unit called the *control* serving exactly this supervisory function. So, just as with the soup bowls, to add two numbers together the clerk picks up the quantities in the two boxes containing the numbers to be added, performs the addition of these items, and places the result in a third box. The operations the clerk follows are what's called an *algorithm,* and they are coded into the program that the machine follows."

Schrödinger interrupted: "This is all extremely interesting, Turing. But I fail to see how a primitive device like a huge post office can do any practically useful computation. In physics, we often need to calculate quantities like the total energy or the angular momentum of a large system of particles, such as atoms or planets. These calculations involve difficult operations, including the adding up of the interactions among thousands—or even millions—of such particles. And even when we do simplified versions of these operations by hand, the computations often generate pages of calculations and equations. How can your 'post-office computer' do this same job better or faster?"

"At first glance, it does seem as if my computer is far too simple to do anything other than the most primitive type of calculation," responded Turing. "But in my paper on computable numbers and Hilbert's Decision Problem, I proved mathematically that any quantity that can be obtained by following a set of rules can be calculated by a 'post-office computer' of exactly the sort I've just described. So our concern is really more with the practical issue of how to actually construct such a machine than it is with the machine's adequacy to the task of calculating this or that quantity."

"Perhaps you could explain just briefly how you managed to establish that a post-office-type of computer can really compute any quantity that can be computed," asked Snow. "This might help us understand both the power and the limitation of these kinds of machines."

"It's really very straightforward," replied Turing. "In my study of computation I made use of a theoretical kind of 'paper computer'. Some people have already taken to calling this a *Turing machine*," he remarked quietly, with a slight touch of pride. "It consists of two elements: an *infinitely long tape*, ruled off into squares, each of which can contain either the symbol '0' or '1', and a *scanning head* that can move forwards or backwards along the tape, one square at a time, reading the symbol on the current square and either leaving it unchanged or writing a new symbol on that square. At any step of the operation, we assume the scanning head can be in one of a finite number of configurations, or what are called *states*. You can think of the machine as having a pointer that at any moment is set at one of the letters A, B, C and so on. This letter then represents the "state" of the machine at that moment. Part of the program tells the machine how to change the pointer setting, depending on what state the machine is currently in and what symbol is on the square of the tape that the head is currently reading."

Reaching into his coat pocket, Turing brought out a note pad upon which he quickly sketched a diagram showing an example of one of these 'post-office' computing machines, whose scanning head could be in any of twelve states, which he labeled A to L.

"The behavior of one of these theoretical machines is fixed by its program, which is a list of instructions telling the machine what to do for any set of circumstances it may happen to find itself in. Since at any stage of operation the scanning head has two pieces of information—the symbol currently being read from

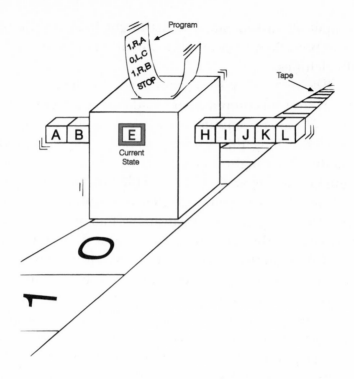

the tape, and its current state—a typical instruction says something like: 'If you're in state A and are reading the symbol 0, move one square to the right and enter, say, the state B.' In this setup, the possible actions the scanning head can take are to:

1. Move left one square.
2. Move right one square.
3. Replace what is written on the current square with a 1.
4. Replace what is written on the current square with a 0.
5. Retain the current state.
6. Change from the current state to another state.
7. STOP."

"All fascinating stuff," interrupted Haldane. "But could you show us how this setup can be used to actually calculate?"

"Of course. What would you like to compute?"

"How about showing the way this gadget would add 1 and 2 to get 3?" replied Snow. "We just want to get the overall idea of its workings on a simple problem."

Turing then went through the steps of how one would use this kind of machine instead of soup bowls and spoons to add the two numbers. "The machine starts with a tape, all of whose squares carry the symbol 0," he explained. "As before, suppose we want to add 1 and 2. We begin by placing a single 1 and a sequence of two 1s on the tape, separating them by a 0 to indicate that they are separate numbers." Then he sketched on his note pad the program for a 3-state machine to perform this addition.[4]

	Symbol Read	
State	1	0
A	1, R, A	1, R, B
B	1, R, B	0, L, C
C	0, STOP	STOP

"Now," he went on, "this program will cause the machine to stop with a sequence of three consecutive 1s on a tape that otherwise contains all 0s. Here's why. By convention, we assume the machine starts in state A,

---

[4]The entries in the table are interpreted as follows: if the scanning head is in state A and reads the symbol 1 from the tape, then the program says that in this situation, the action of the scanning head is (1, R, A). This is Turing-machine shorthand for telling the head to "replace the symbol on the current square with a 1, move one square to the R(ight) and enter state A". The other instructions in the program are interpreted similarly.

reading the first non-zero symbol on the left. Since this symbol is necessarily a 1, the program tells the machine to re-write the 1 on the square, move one square to the right and remain in state A. Since the machine is still in state A, and the current symbol read is again a 1, the machine repeats the previous step, moving one square further to the right. But now the scanning head reads a 0. The program says that in this case the machine should print a 1, move to the right and enter state B. Carrying out the remaining steps of the program, the scanning head finally stops with the tape looking just like the input tape—except that the 0 separating the sequences of 1s representing the numbers 1 and 2 has been eliminated. So, just as I claimed, the tape now has three 1s in a row, which is the result of adding the two numbers originally placed on the tape. And, of course, you can see that there's nothing special about the numbers 1 and 2. The very same program could be used to add *any* two numbers."

Again Schrödinger interrupted, complaining that "This computing machine looks fine for adding small numbers like 1 and 2. But it doesn't look very promising as a practical device for carrying out more extensive calculations like those done in physics. For instance, adding 1,234,567,890 and 9,876,543,210, or computing the square root of π, would seem to take a very long time to do in such a fashion. Just to express these numbers on your input tape would probably require a tape several hundred feet long."

"It all depends on what you mean by *practical,*" countered Turing. "If you have to do the operations by hand, as I did here in showing you how to add 1 and 2, then it's a pretty slow business and I would agree that it doesn't seem very practical. But if you can arrange for electro-mechanical devices to do things like moving the tape

scanning head and carrying out the steps of the program at speeds far faster than any human being is capable of, then we open up entirely new possibilities for
such machines to carry out calculations that no human
could ever perform. "

Here Turing paused for a moment, thinking of
his old schoolmate David Champernowne, who had
once posed the same kind of question as Schrödinger
about doing calculations with large numbers. Champernowne had even invented the special number 1,234,
567,891,011,121,314, ..., formed by writing down the
positive integers in order, as a kind of test case for
what could be computed on Turing's machine. Turing
smiled to himself when he thought of how they had
labeled this number, 'The All-Time Champ' in Champernowne's honor. His attention was jerked back to the
present when Snow asked: "Is this what you meant earlier when you said that such a primitive-looking device
is capable of doing any kind of calculation that could
be conceived of?"

"Exactly. My 1936 paper showed that while it may
take a large number of steps or a very long tape for
this kind of computing machine to carry out the types
of computations Schrödinger refers to, *anything* at all
that can be thought of as the result of following a set
of rules—including every possible type of numerical
computation—can be calculated in a similar step-by-
step fashion by this kind of machine. The only conditions are that you have enough time to go through
all the steps of the calculation, and that you have a
long enough tape to store all the intermediate results.
Strictly speaking, my theoretical computer must have
an infinite number of tape squares, or 'postal boxes' at
its disposal."

"Ah!" shot back Haldane. "I thought there must be

a catch somewhere. When was the last time any of you saw an infinite post office. Sounds just like the kind of fantasy world mathematicians are notorious for dreaming up. What does this have to do with real-world calculations like the ones Schrödinger just mentioned?"

Turing explained, "What the machine really needs is the ability to add as many tape squares as might be required to store the intermediate results of any particular calculation. So for any calculation that stops after a finite number of steps, the machine doesn't really need an infinite tape, after all, but only a finite one."

Picking up on this point, Snow raised the question: "Can you tell beforehand how long a tape will be needed for any particular calculation?"

"No, you can't," replied Turing somewhat defensively. "In general, there's just no way to tell before you start a calculation how long a tape you're going to need or how many steps it will take to finish. You just have to be ready to expand the length of the tape as the calculation unfolds."

He went on to suggest that "If a particular program and the input data the program is to process are given, it would certainly be useful to know before starting out on the computation how many steps and how many squares on the tape will be required to complete the calculation. Ideally, one would like to have a kind of 'metaprogram' that would accept the original program and input data as *its* input data, and then work out how much time and tape storage space the original program will need in order to complete its task. As a result of Gödel's work, I was already suspicious in 1935 about the existence of any type of metaprogram that would solve this *halting problem*. And in the same paper in which I introduced my hypothetical computing machine, I established the fact that it is indeed impossible to ever find such a 'universal' procedure for determining how many squares on the tape a particular calculation requires."

"So," observed Schrödinger, "these results by Gödel show that there exists a problem that cannot be solved by following the steps of a program. Your results, on the other hand, say that for a *given* problem, there is no way to know if it is solvable or not."

Turing nodded in affirmation—and admiration— as he had never really thought of the relationship of his own work to that of Gödel's in quite this way before.

"But this result doesn't mean that such a decision procedure can't be found for some programs and for some input data, does it?" continued Schrödinger. "For example, a program whose job it is to read the input tape and stop when it comes to the first 1 surely has such a procedure, since in that case there is the simple decision rule: if the input data contains even a single 1, then the program will halt after a finite number of steps, the exact number determined by how far the first 1 is away from the starting square; otherwise, the program will run on forever. So all you have to do is look at the input tape and see if there's a 1 on it. If there is, the program will halt; otherwise, it won't."

"Of course," responded Turing. "The difficulty lies in producing a single solution to the halting problem that works in *every* case—for every possible program and set of data. And most programs are a lot more complicated than yours, having a stopping criterion that usually depends on quantities that are produced during the course of the computation. Such stopping rules take the form: if this or that quantity appears, stop; otherwise, keep calculating. But we can't generally tell what kinds of quantities will turn up until we actually do the computation."

"All this talk about computing machinery, infinite tapes, stopping criteria and the like still doesn't sound very useful to me," grumbled Haldane. "It's just making me hungry. I agree with Schrödinger that Turing's paper computer might well be able to *theoretically* calculate anything that can be calculated by following a set of rules, or a 'program'. But what I'd like to know is what is such a machine good for? Can this *theory* of computing be turned into an actual device for performing real-world computations of the sort involved in things like building bridges or running trains?"

"Practical computations for building bridges are one thing," said Wittgenstein, rousing himself to finally enter the conversation. "Perhaps Turing will someday be able to build a mechanical computer that will gladden the heart of a physicist and brighten the day of an engineer. But as Snow has reminded us, we're here to discuss a much deeper issue, the problem of whether one of these machines can actually *think* like a human being. And I don't begin to see how writing and erasing a lot of 0s and 1s on a long tape has anything at all to do with thinking. How can anyone believe for a single moment that this sort of symbol-writing machine bears any resemblance to the thought processes going on in a human brain? Brains are not machines, and it would be a categorical mistake to belive that they are."

Turing begged to differ: "First, let me explain about the physical makeup of the brain. I think you'll see then how its structure is captured in the structure of a computing machine. Perhaps then we can discuss more fruitfully how such a device might actually be capable of true thought."

Sucking in his breath, Turing paused for a moment, thinking back to his many readings and conversations on neurophysiology, and biological processes. Finally, he offered the group a greatly compressed and simplified account of how the brain works.

"The human brain is composed of a very large number of elements called neurons, ten thousand million or more according to some people's estimates. These neurons are connected to each other through an extremely dense network of 'wires' called axons and dendrites."

"Something like a giant telephone switching network, is that it, Turing?" asked Haldane.

"That's a good image to keep in mind," agreed Turing. He then continued. "The neuron itself can be thought of as a particularly primitive kind of switch that at any moment can be only either 'ON' or 'OFF'. Which of these is the case is determined by the signals the neuron receives from the other neurons to which it's connected."

"So it's a bit like shooting a gun," interrupted Haldane. "You exert pressure on the trigger, and when the pressure exceeds the strength of the spring in the trigger, the gun fires. The only difference is that each neuron may have many signals coming to it, signals that are themselves the outputs from other neurons. But even though it may have many inputs, an individual neuron has only a single output going to another neuron. Some of a neuron's input channels are *excitatory*, which is like putting more pressure on the trigger; other input channels are *inhibitory*, so that signals coming into the neuron along these pathways subtract from the neuron's overall input, like reducing the pressure on the trigger. If the sum of all these positive and negative inputs exceeds a particular threshold level, the neuron fires a pulse at its output channel; otherwise, it remains OFF."

"If I may continue?" said Turing, with some asperity. "Most neurophysiologists and psychologists believe that the patterns created in the brain by these neuronal firings form an important part of the basis of human thought processes and, hence, human behavior."

As Turing paused, Schrödinger stepped into the fray. "I'm now starting to see how you might make an analogy between a computing machine and a brain. They both involve the storage of large numbers of elementary data, a 0 or a 1 on a square of the computer's tape, an ON or an OFF state in a neuron. Moreover, both the computer and the brain process these data into patterns. Is this the basic analogy you're pursuing, Turing?"

"Precisely. The brain stores its data in the form of patterns created by the firing of its neurons. Each such pattern is just a listing of which neurons are ON and which are OFF at any given moment. These patterns are then associated with what we call 'thoughts' in ways that no one yet really understands. The computer, on the other hand, stores its data in the 'postal boxes' that I spoke of a few moments ago. This pattern, too, is simply a sequence of 0s and 1s, or what's the same thing, ONs and OFFs. And in both cases there's a way to modify what's stored in an individual memory location, either by causing different neurons to fire in the brain or by executing an instruction from the machine's program. It's the striking similarity between the functional activities of storage and pattern changes in the brain's neurons and the same activities in the workings of the computing machine that leads me to believe that we can actually build a thinking machine. The only obstacle looks to be technological, not logical."

Snow then added: "The key element here, though, is what these brain components *do*, not what they are made of. And that means we have to look at what is happening in the brain's cortex, which is where higher human cognition seems to take place."

"Exactly so," Haldane interjected, explaining that "The cortex is a continuously folding layer forming the outside of the brain. In humans this region is often termed the *neocortex*, and is the newest part of the brain, evolutionarily speaking. More importantly, it's the part of the brain where reasoning and thought occur. The cortex can be divided into a great many areas, both structurally and functionally. But all the parts are built of the same basic components and are linked together in similar ways. So the various functions associated with different parts of the cortex are probably due to the different sensory signals coming into them, not to a difference of structure."

Turing then described some pathbreaking results obtained a few years earlier that gave the theoretical basis for his brain-machine analogy. "In 1943," he said, "Warren McCulloch, a neurophysiologist at the University of Illinois, and Walter Pitts, a student in maths at the University of Chicago, published a marvelous article about how the operation of a group of neurons connected with other neurons might be duplicated using purely logical elements. The model regards a neuron as being activated and then firing another neuron in the same way that a proposition in a logical sequence can imply the truth or falsity of some other proposition. Moreover, we can picture the analogy between neurons and logic in engineering terms as signals that either pass—or fail to pass—through an electrical circuit. It's only a small step, at least in principle, from the abstract logical structure developed by McCulloch and Pitts to its implementation in the physical elements of an electronic computing machine."

Wittgenstein could contain himself no longer. Tossing down his napkin, he leaned across the table to challenge Turing's claims. "Surely you're not saying that the pattern of data stored in these various postal boxes in the machine or in the ON/OFF pattern of neurons in the brain can be interpreted as thoughts! Suppose you smell fresh bread baking or have a picture in your mind of your grandmother's face. Now if I open up your skull and look at all those neurons in your brain going ON and OFF, I certainly would not be able to say: 'Ah, here's pattern A, so Turing must be thinking about a slice of fresh bread. And now here comes pattern B, so Turing has changed his mind and is now thinking about a visit to his grandmother's house.' "

Pausing only for a moment to catch his breath, Wittgenstein went on.

"I can't observe the mental phenomena of others. And I can't observe my own either, in the proper sense of 'observe'. So where are we? In a fog, that's where. We're in a set of confusions that can't be resolved by introspection or by behavioral analysis. Nor can it be resolved by a *theory* of thinking, either. The only resolution comes from a conceptual investigation, an analysis of how we make use of words like 'intention', 'will' and 'hope'. These words gain their meaning from a form of life, a language game, quite different from that of describing and explaining the ordinary physical phenomena of daily life."

"And," he continued, "the same must be true for a computing machine. If I take the cover off the machine and look at each of the squares on its tape, the pattern of data formed by the symbols written on these squares surely can't tell me what the machine is thinking about. In fact, I can't see how you can say that it's 'thinking', at all. It takes a human being standing outside the machine to interpret these patterns as being *about* something."

Schrödinger interrupted this tirade, asking, "Are you denying that there are laws of thought that we can uncover to explain thinking, just like we use the law of gravity or the laws of chemistry to explain physical phenomena?"

"I'm saying that the whole modern conception of the world is founded on the illusion that the so-called 'laws of nature' are the explanation of natural phenomena," Wittgenstein replied.

Snow broke in to stop this argument, at least for the moment. "This is going to require a bit of an explanation of its own, Wittgenstein, if you're implying that not only does human thought transcend the following of rules, but so do all other natural processes. But I see that Simmons is ready to serve the fish. So I suggest we pause for a few minutes and fill our glasses with a bit of this fine-looking Montrachet while he does his duty."

Chapter Three

The Fish

Minds and Machines

Simmons moved around the table, serving each of the guests a lightly broiled sole meunière swimming in butter, as Snow reflected on the puzzle laid out by Wittgenstein. How is it that a collection of 1s and 0s on a tape, or for that matter a pattern of ONs and OFFs in the neurons of a brain can give rise to thoughts? How is it possible that a mere collection of more-or-less arbitrary symbols written on a tape—or stored in a brain—can come to *mean* something as disparate as the sound of a church bell, the flash of a bolt of lightning or even this very puzzle I'm worrying at involving the relationship between symbols and thought, he wondered Surely there must be more to thinking than just changing a string of 0s and 1s into another such string—no matter how quickly the transformation takes place or how many of these strings can be changed at once. Turing can't really be serious, can he, when he says that a machine doing no more than this kind of symbol shuffling is capable of actually *duplicating* the thought processes of the human mind?

Giving voice to these concerns, Snow turned to Turing and said, "I think I can speak for many of us when I admit that I find it literally incredible to think that a machine that can only move 0s and 1s about on a tape is capable of human thought. Perhaps it would help us more easily capture the essence of your argument if you would explain *exactly* how you think these strings of symbols in your machine actually come to mean something."

Looking up for a moment from the delicacies on his plate, Haldane seconded Snow's request. "I'm also bothered by the idea of how the purely syntactic manipulations of the symbol strings on a computing machine's tape could ever give rise to rich, semantically laden objects like this excellent piece of sole sitting on this plate. Tell me this, Turing, where should I look on your tape to find this delectable piece of fish, eh?" he asked, holding up a morsel of fish on his fork. "Answer me that, if you will, and I'll concede that your machine just might be capable of having thoughts like mine."

Seeing the difficulties he faced in trying to explain exactly how this syntax-to-semantic transmogrification might possibly occur, Turing stared out for a moment at the rain that was by now beating harder than ever against the window panes. Fidgeting about in his chair, he was somewhat taken aback by both Snow's puzzlement and the intensity of Haldane's query. How can anyone *explain* scientifically a gut feeling or a firmly held conviction? he asked himself. What kind of logical arguments can I give that would convince a hard-nosed materialist like Haldane or Schrödinger that intelligence is simply a matter of following the right kind of rules? Whatever I say, Wittgenstein will surely argue to his last breath against me. Why did I ever agree to come to this gathering tonight? The situation is truly hopeless. But I'm in it now up to my neck, so I suppose there's nothing for it but to jump in and hope for the best.

Taking a long draught of water and clearing his throat, Turing launched into an explanation of how symbols on a tape might give rise to genuine thought. First, he told the group how any type of idea, object, or action that can be expressed in language can be *coded* by a string of 0s and 1s on the tape of a computing machine.

"Let's talk about Haldane's piece of sole, for just a moment," he began. "Suppose I want to represent the word SOLE on the tape of my machine. One simple way of doing this is to set up a scheme in which every symbol in the Latin alphabet has its own unique string of 0s and 1s. There are many ways to do this type of coding, but let me just show you one of them. Take a block of eight tape squares, each of which can contain a 0 or a 1. There is then a total of $2 \times 2 \times 2 \times \cdots \times 2 = 2^8 = 256$ distinct patterns of 0s and 1s that this set of eight squares can display. So I can associate each of these patterns with one of 256 symbols, which is quite sufficient to accommodate all of the letters and symbols that one might find in, say, a large dictionary of the English language. For instance, suppose I decide to let the string 00000001 represent the lower-case letter a, while assigning the pattern 00000010 to the letter b, and so on. In this fashion, each letter of the alphabet, the numerals 0 to 9, punctuation marks like ? and !, as well as other orthographic symbols such as (and], can be given its own unique code as a sequence of eight 0s and 1s. Using this scheme, I can then represent the word SOLE on the tape by a set of four groups of eight squares—one group for each of the four letters in the word. And by putting in the appropriate punctuation and other symbols of written English, the same scheme will allow me to code on the tape any notion whatsoever that can be communicated in written form."

"This is all perfectly clear," said Schrödinger. "But it simply trades one set of symbols for another."

"True," agreed Turing. "But once we have coded a particular thought or situation into a string of 0s and 1s in this way, the program of the machine can then transform the strings into new strings. And these new strings can then be *decoded* into statements in English, one of which might express the texture and taste of the piece of sole on Haldane's fork or even the thoughts running through his mind as he contemplates the pleasure he'll receive when he bites into it."

"Are you saying that the rules of the program changing the symbols on the tape are doing the same thing that the human brain does when it turns neurons ON and OFF in the process of thinking?" enquired Snow, eyebrows raised incredulously.

"Basically, yes. Of course, we don't really know yet what those rules are that the brain uses. For that matter, we don't even know *how* the brain actually stores and uses its rules.

"A merely passive transformation of symbols by a predefined rule is not my idea of thinking," butted-in Haldane. "But perhaps if the program has the capacity to 'learn' by examining the results of its actions and modifying them accordingly, then that might be enough that we would be tempted to call such a program 'intelligent'."

Regaining the floor, Turing continued his description of how he felt the brain worked to create thought. "I'm convinced that what a computing machine does in changing the symbols on the tape into new symbols is exactly the same kind of process that the brain goes through in the process of thinking when it causes different patterns of neurons to fire at different times and thereby generate what we call 'thoughts'."

"Sheer sophistry," cried Wittgenstein. "Where do you find the *meaning* of the word SOLE in all this symbolism? How can you claim that a string of symbols like 0s and 1s put together according to a completely arbitrary coding scheme could in any manner actually *mean* that slice of fish sitting there at the end of Haldane's fork? The naming of the piece of protein we call 'fillet of sole' can only take place within the context of a developed language, one in which there already exist rules for picking out objects, using names and doing operations. The criteria for this are not in the logic of machines, tapes and codes, but in the actual practice of a language community. You cannot infuse this kind of meaning into a dead string of symbols simply by making up a set of rules saying how to transform these strings into new ones."

"Are you telling us that meaning arises only out of the kind of social consensus that allows us to communicate with each other by natural language?" asked Snow in a rather surprised tone of voice.

"Exactly. Meaning can only come from participation in a language game. Computing machines can never be players in the kind of game we're playing right now. Turing is quite mistaken to believe that what the machine might 'think' of as a piece of sole bears any resemblance to what any of us here think about that same morsel of fish. We have a common belief about that piece of fish because we have a shared way of life. And if the machine is playing any language game at all, which I seriously doubt, it certainly isn't any of the games played by human beings. In the final analysis, meaning resides in social practice, not in logic."

As Wittgenstein became increasingly animated and agitated in expressing his point, Schrödinger leaned over and placed his hand on his countryman's arm, attempting to calm him down a bit. "Just a moment, Wittgenstein," he said. "Turing may have a point here. Even if the kind of meaning we would attach to Haldane's delectable morsel of sole does come from human experience and participation in a shared way of life, it's not at all clear to me that at least in principle this meaning couldn't be coded into the neural circuitry of our brains in just the way that Turing claims that it can be coded on to the tape of a computing machine. What I don't see in this line of argument, though, is how the kind of intelligent behavior associated with learning enters into Turing's scheme of things."

"Yes," chimed in Snow. "As Turing has just described it, there doesn't appear to be any way that his machine can do anything other than move the symbols around on the tape in accordance with the *predefined* instructions making up the machine's program. But this is not at all the way human beings behave. We're always ready to change our minds, adapt to new circumstances, give inconsistent responses in seemingly identical situations and in general behave in strange and unpredictable ways. If a computing machine can't do this, then I don't see how it could ever display anything like what we would call human intelligence."

Agreeing with both Snow and Schrödinger, Turing responded: "A computing machine will be able to display intelligence only if it is able to modify its program in the light of new information. So as the machine reads new input patterns presented on its tape, it will need to have rules for changing its current rules of operation, 'meta-rules', if you like. In this way, the program could learn and adapt—just like human beings do—to a changing environment and to circumstances that it 'sees' via the inputs on the tape."

"But in order to do this you would have to give the machine the very same sensory inputs we all have and allow it to 'grow up', so to speak, in the same environment as a human, wouldn't you?" asked Schrödinger quietly.

Turing thought for a moment, and then replied: "Let me slightly restate my case for the logical possibility of a thinking machine. An important reason underlying my belief in the feasibility of the idea is the fact that it is possible to make machinery to imitate almost any small part of a man. The microphone does this for the ear, and the camera does the same for the eye. The questions we are discussing here are concerned primarily with the nervous system. And we could certainly build fairly accurate electrical models to copy the structure and the behavior of nerves, although there seems very little point in actually doing so. It would be rather like putting a lot of work into cars that walked on legs instead of continuing to use wheels."

"Are you really suggesting that to build a 'thinking machine' one should take a man as a whole, and try to replace each part of him with machinery?" challenged Haldane. "This would surely be a monumental undertaking, and even if the task could be accomplished the creature would still have no contact with food, sex, sport and many other things of considerable interest to human beings."

"Although this is probably the 'sure' way of producing a thinking machine, it seems to be altogether too slow and impracticable," Turing replied. "My proposal instead is to try and see what can be done with a 'brain' that is more or less without a body, providing it with, at most, organs of sight, speech and hearing.

Of course, we are then faced with the problem of finding suitable branches of thought for the machine to exercise its powers in. In this connection, I think areas like chess-playing, cryptography and mathematics are good candidates, since they require little contact with the outside world."

"Let's assume for a moment you could give your machine these sensory inputs," said Haldane. "How would it be able to use them to actually change its internal programming so as to learn about even such a restricted domain as mathematics or chess?"

"Well," speculated Turing, "we should probably begin with a machine with very little capacity to carry out elaborate operations or to react in a disciplined manner to orders. Then, by applying appropriate interference that mimics education, we should hope to modify the machine until it could be relied upon to produce definite reactions to certain commands."

"What about a very restricted domain of discourse like mathematics," said Wittgenstein. "How would a machine learn maths?"

"In the case of mathematics, this would involve telling the machine about sets of objects, like points and lines, as well as about the logical operations needed to form new sets. But I can't really give you a full and detailed account of exactly how to do this now, for the simple reason that I don't know how to do it—yet! This is the main thrust of our current research effort. But I am fully convinced that there is no logical or technological barrier to carrying out this plan. What's lacking at the moment is only the will—and, of course, the resources—to do it."

Haldane then noted that "A moment ago you told us about the work by McCulloch and Pitts attempting to mimic the neuronal circuits in the brain by mathemat-

ical formulae that could, at least in principle, be built
from modern electrical components like valves, relays
and the like. Maybe a scheme for adjusting the strength
of the connections between the artificial neurons in
such circuitry could serve to give such a machine the
capacity to learn and adapt."

"My thoughts precisely," replied Turing. "Let me
show you how one of these artificial neuronal nets might
look."

Reaching for his notebook again, Turing sketched a
diagram showing how a network consisting of the ideal-
ized neurons envisioned by McCulloch and Pitts might
be structured.

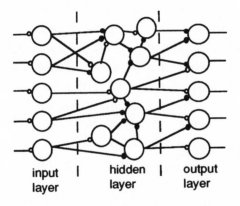

input   hidden   output
layer   layer   layer

"I am saying that by regarding the ON/OFF pattern
of the input neurons as being analogous to the input
symbols on the machine's tape, while letting the firing
pattern of the output neurons correspond to the output symbols on the tape, the connections among the
neurons would transform the pattern on the input layer
into a pattern at the output layer. This means that these
connections perform the very same function as the program of the computing machine. So the two systems—a
neural network and a Turing machine—are completely
equivalent. What one can do, so can the other."

"Does this mean that McCulloch and Pitts proved there is no difference between a network of mathematical neurons like this and the abstract computer you showed us earlier?" asked Snow.

"Yes! That is precisely what they showed," answered Turing, breaking into a stammer in his enthusiasm for describing this work. "Both the neuronal net and the machine perform exactly the same operations; in mathematical jargon, they are 'isomorphic'. So whatever can be done by one of these neural networks can be done by one of my computing machines, and vice versa. So, for instance, Haldane's mental state when he's thinking about that piece of sole on his fork can be regarded as nothing more than a stage in a computing machine's program. And since we can actually construct both neural networks and computing machines using electronic devices, we can regard these devices as providing an electro-mechanical theory of mental states. In a sense, this means that biology equals electronics."

"You've been rather silent, Schrödinger," observed Snow. "As a physicist, what do you think about the idea of building an artificial brain from valves, wires and such?"

"What's puzzling me at the moment," replied Schrödinger, "is whether Turing thinks that we can build an electronic machine to *mimic* the human brain in some of its functions? Or is he saying that it's possible to actually *duplicate* or faithfully *reproduce* a human brain in electronic form? I wonder, Turing, if you could clear up this point?"

"I'll try. From the standpoint of displaying human-like intelligent behavior, I don't see that it really makes much difference. Unless, of course, you think there's something special about the material constitution of the human brain that accounts for its cognitive abilities, and that this 'something special' cannot be captured in electronic circuitry."

"Well, it won't take a fortune-teller for us to guess your view on this, Turing," said Snow with a smile. "But why don't you tell us, anyway?"

"Indeed," replied Turing. "My feelings are clear and definite on this matter. I believe that there is nothing at all special about the material composition of our brains, at least insofar as thought is concerned."

"You might even say that matter doesn't really matter, eh?" joked Haldane.

"That's certainly one way to put it," Turing replied, with a slight hint of a smile. "What does matter, though, is what the brain's components—basically its neurons— actually do, and the way they are connected to each other. It is these functional and structural aspects of the brain that give it its cognitive power. I'm convinced that if we built electronic neurons and connected them as they are connected in the human brain, then that electronic device would embody rules for thought and action of precisely the same sort as those really present in the human brain. Such a machine would carry out exactly the same functions that the brain does. Machine intelligence comes from the complexity of the totality of rules constituting the program, not from the individual steps of the program, which may be very elementary indeed, as we have already seen when adding 1 and 2."

Snow saw that with these rather daring assertions, Turing was making an identification between the continuum picture of learning, information processing and cognitive activity advocated by the behavioral psychologists and the mechanism of computation.

"As I see it, by substituting the notion of a network of mechanical rules for the behaviorist's causal network of stimulus-response connections, you're claiming that one can capture the logical grammar of purpose, choice and learning within the framework of a set of mechanical rules that could be encapsulated in the instructions of a computer program," added Schrödinger.

Jumping up from the table and pushing back his chair, Wittgenstein could contain himself no longer. Pacing back and forth across the room, his eyes gazing off into a realm beyond space and time that only he could see, Wittgenstein challenged Turing.

"Under what circumstances can I say that someone is following a rule? If you push the buttons '20', '25' and '×' on a calculator, arriving at the number 500, this does *not* mean that you have calculated 20 × 25. The question: 'How did you arrive at the correct answer?' is a question about the rules that were used. Just coming up with the correct answer does not allow us to say that someone—or something—is calculating."

"But Turing is saying that that is exactly all it takes; it is only the behavior that matters, not how the behavior was arrived at," noted Schrödinger.

"But I say that if calculating looks to us like the action of a machine," continued Wittgenstein, glaring over at Schrödinger for having the temerity to interrupt him, "it is the human being doing the calculation that is the machine. Just because a rule can be mechanized does not mean the rule is 'mechanical'. Any rule can be regarded as a description of a mechanism. Turing machines are simply humans who calculate. The question of whether a machine can think is just plain unanswerable because it is logically absurd. It's like asking what is the color of 3."

Then, like a slowly leaking balloon, Wittgenstein seemed to run out of gas. With a pained and rather distracted look on his face, he returned to his chair, slumped over and looked down at the table, seeming to lose track of the discussion entirely.

Meanwhile, Haldane turned to Turing saying, "Wittgenstein's argument certainly seems to suggest there's much more to thinking than following a set of rules. I wonder how one could ever determine whether one of your 'electronic brains' was really thinking, or was just producing results from a set of rules that made it appear *as if* it were thinking like you and me. Is there any objective test one might use to separate these possibilities?"

"How does one ever decide whether another human is thinking?" replied Turing, rather tetchily. "None of us has access to anyone else's inner mental life. All we can do is judge on the basis of how a person behaves. I say something or do something to you, and you react in a certain way. I then react to your response and so on. After a sequence of such interactions, I decide that you are a thinking being instead of a lump of inanimate matter like this water pitcher or the knife on my plate. That's the way we come to see other humans as thinking like ourselves."

"How would you ever empirically test this sort of 'thinking'?" asked Snow.

"I would propose the following kind of test. Let me place a computing machine programmed to think like a human and a real human being side by side in the next room, giving each of them a teletype connection to a typewriter sitting in this room. I then ask you to sit down at the typewriter and carry on a written conversation with one or the other of them via the teletype—but I don't tell you whether it's the machine or the human that you are conversing with."

"What kind of questions can I ask?" enquired Snow.

"You can ask any kind of question you like, make

any type of statement and, in general, converse back and forth just as we have been doing here at the table tonight."

Ah, thought Schrödinger, Turing is setting up a kind of *gedanken* experiment to illustrate the notion of thought, reminding him of his own, now-famous, thought experiment involving a closed box and a cat that he had invented to illustrate some of the puzzles surrounding the act of measurement in quantum theory.

"Now suppose I allow you to interact in this fashion with whatever or whoever is in the next room," continued Turing, "let us say you interact for one hour and that we perform this experiment many times. If at the end of this set of experiments you are unable to reliably distinguish the human from the machine, then I would argue that either the machine is intelligent or you, the human, are not. So if you're willing to accept humans as being intelligent, then I don't see how you can fail to accord the same rights to the machine. After all, this procedure is exactly how I am deciding at this very moment that you are indeed intelligent. By observing your reactions to what I am saying and doing in a variety of circumstances, I have concluded that you are a thinking being like myself. And it's not because you have a moustache or two eyes or any other reason pertaining to your physical appearance. It's solely because you act and react in certain ways that I accept as being the normal reactions of intelligent humans in such circumstances."

"So you're saying that this kind of 'Imitation Game' is the right sort of objective test to decide if a computing machine is capable of thinking like a human. Is that your argument?" asked Snow.

"Indeed it is," replied Turing.

"By focusing your test for intelligence on the external behavior of a machine or a person rather than on what's taking place within that machine or person's brain, you've placed yourself directly at the center of the behaviorist tradition in psychology," noted Schrödinger. He recalled that while being rather ambivalent in their theories as to the role of internal 'mental states' acting as causes of thought, behaviorists in the 1920s, people like John D. Watson, were adamant in denying that any such internal characteristics of the brain could be allowed to contribute to a *scientific* theory of human thought processes or behavior.

"According to these so-called behaviorists, it is only the externally observed actions that can form the basis for a legitimate scientific theory of behavior," added Haldane. "Turing's test for intelligence, then, seems to be simply a transference of this behavioral paradigm from man to machine."

As with the arguments of the behaviorists themselves, Turing's test came under immediate attack from almost everyone around the table.

Haldane began the assault on the Imitation Game by arguing, "It does seem that one tends to learn more about the nature of another being by fighting it than by obeying it. But mechanisms cannot feel pleasure, be warmed by flattery or in general show any kind of conscious emotional reactions. So it seems to me that the only way we could be sure that a machine thinks is to actually *be* the machine."

"That's a very solipsistic view that would make the communication of ideas impossible," Turing replied. "If you were right—which I don't for a moment think you are—then my Imitation Game certainly wouldn't suffice as a test for intelligence, since we could never be sure anyone else is really thinking without actually *being* that other person. Yet I'm perfectly willing to concede that you are thinking, and I suspect that you'd be ready to admit the same about me. So I'm sorry to have to say that I find this argument quite unconvincing. Solipsism is not an answer to anything."

Snow then returned to the fray, saying, "Perhaps so, Turing. But from what you've said so far about the workings of the computing machine, it seems to me that the machine can only do what we order it to do. It has a particular set of instructions that constitute its program, and these instructions are slavishly followed, step by step, until the machine stops. So I don't see how the machine could ever display unpredictability, free will, inconsistency or any of the many other things we see in everyday human behavior."

Turing immediately responded: "This is really the same kind of objection that Lord Byron's daughter, Lady Lovelace, put forward nearly a century ago when she was working with Charles Babbage on his 'analytical engine'. And I'll say the same thing to you that Babbage probably said to her. It's just not always clear what consequences follow from a given fact; in particular, it's very unclear what kind of quantities will be computed over the course of a computation that's being carried out by following a given set of rules. Even if the rules are simple when taken individually, going through a succession of many thousands—or thousands of millions—of steps using such rules can easily generate entirely unexpected quantities. A machine that learns, and is able to modify the instructions in its program on the basis of new inputs that are presented, serves as an excellent example of such a rule-following mechanism. So yes, the machine may well be able only to do what we instruct it to do. But even we can't foresee the consequences of those instructions."

At this point Schrödinger remarked, "Wittgenstein has already given some excellent reasons in support of the informality of human behavior. It would seem that it's just not possible to provide rules of conduct to cover every eventuality; in short, there looks to be a lot more to life than merely following a set of rules. So I don't see how any type of machine—even one capable of modifying its rules—can ever duplicate human behavioral patterns if some of those patterns are not governed by *any* rule at all."

Turing was a bit puzzled by such an objection, especially coming from a scientist of Schrödinger's stature, since he thought it was tantamount to denying the existence of structure or pattern in human behavior. Replying to Schrödinger, he said, "the only way to discover laws of behavior is to search for them. But we can never be sure we have looked hard enough and long enough. Perhaps behaviors that we think lie outside the domain of rules really are determined by some rules or other, and we have just not been clever enough or diligent enough in our search for them."

Schrödinger parried this reply by bringing up Gödel's work in mathematical logic, which Turing had mentioned earlier. "But you yourself told us that Gödel demonstrated that there exist statements about numbers that can never be proved or disproved by following a set of logical rules. Yet we humans can see that such statements must necessarily be true; we just can't *prove* them to be so. Doesn't this mean that there are things that the human mind can know that can never be known by any machine?"

"Gödel's results are rather a red herring in this context," Turing objected. "Gödel's Incompleteness Theorem assumes that the logical system one uses to prove or disprove statements about numbers is consistent and error-free. This means that it is not possible to prove and disprove the same statement within the rules of the system, and that we never make a logical mistake when we apply the deductive rules of the system. If either of these conditions fails, then so do Gödel's conclusions. But human beings *do* make mistakes and they *do* act inconsistently. And a machine that duplicates human behavior patterns would have to do the same. So I don't see how Gödel's results really apply here."

Taking a break from the discussion, which by now had become somewhat heated, Turing reached for the water pitcher while the others debated the claims and counterclaims that had been flying fast and furious across the table.

Snow rekindled the discussion by attempting to summarize the situation thus far: "It seems to me that the crux of the competing arguments revolves about whether a set of rules alone can ever serve as the sole generators of human behavior, specifically cognitive behavior. Turing's case hinges on the assumption that a set of rules, if extensive enough or if allowed to act over a sufficiently large number of steps, can lead to behavior that looks logical, spontaneous, emotional, irrational and/or creative to an outside observer. Wittgenstein's counterargument is that no amount of rule-following, however lengthy or complicated, can ever account for the fullness of human cognitive life. At best, such a rule-following machine could only mimic or simulate a small part of the human experience. So the only way to duplicate a human is to *be* one."

"The idea of a thinking machine is just too horrible to contemplate," exclaimed Haldane, holding out his glass to allow Snow to pour him a little more of the Montrachet. "Some of our more theologically oriented colleagues here in Cambridge might tell you that thinking is a function of man's immortal soul. Machines do not have souls. Therefore, machines cannot think—ever! What do you say to that, Turing?"

"Don't you feel that line of argument implies a serious restriction on the omnipotency of the Almighty?" countered Turing. "You seem to want to believe that humans are in some subtle way superior to the rest of creation. If that's so, then of course we will all have to accede to this fact and I will happily abandon my vision of a machine that can think like a human. But to the best of my knowledge there is no such proof of the intrinsic superiority of humans. So until someone produces an airtight argument for this, I'm forced to regard this type of objection as just so much fanciful and wishful thinking."

Haldane then threw in what sounded to everyone like a wildly speculative, almost half-joking possibility for defeating the Imitation Game. "Suppose," he said, "your interrogator had an extrasensory communication channel with whoever or whatever was on the other side of the screen, so that he could distinguish the machine from the human without using the teletype. Wouldn't that nullify your test?"

"Good heavens, man," exclaimed Wittgenstein, "you're introducing something even more speculative than Turing's cognitive machine. If we're going to let this discussion take off into the mystical stratosphere, why not consider a divine inspiration telling the interrogator what's on the other side of the screen? You've really gone a bit too far with this ESP idea."

Surprisingly, Turing remained silent for a moment or two before responding to Haldane's ESP objection. "I should think that extrasensory communication of this sort would indeed invalidate my test for intelligence. All I can say is that if you admit ESP, then anything might happen. In that case the Imitation Game would no longer be a good way to tell if the machine was thinking like a human or not, since one could then always pick out the human from the machine. But until the existence of such types of communication is scientifically established, I will continue to support the Imitation Game as the right way to proceed."

During this interchange, Snow was silently reflecting on Wittgenstein's earlier argument about the social basis for language. Suddenly, he saw how it all fitted together with his long-held belief that words are always simpler than the reality they represent; for if not, he felt discussion and collective action would be impossible. So if Wittgenstein is right in saying that words somehow emerge out of the social collective, thought Snow, then it would make sense for that collective to agree on expressions for brute reality that are simpler than the reality itself. Otherwise, language could never serve as a kind of shorthand for communication. Pleased by this glimmer of insight into the relationship between language and the world, Snow's concentration was broken by the appearance of Simmons at the doorway, enquiring if he might serve the main course.

"Simmons appears ready to serve. So perhaps this is a good time to break off this discussion for a moment and enjoy a brief pause before the main course," Snow suggested to the group.

"Indeed," responded Schrödinger. "Turing has yet again given us something to think about with this Imitation Game, since it seems to relate directly to the problem Wittgenstein set before us earlier involving the relationship between thought and language. Perhaps we could venture just a little bit deeper into this matter and its relationship to the thinking-machine problem. To my eye, at least, if we employ the test Turing suggests to determine whether or not a machine is indeed thinking, then I don't see how we can avoid coming to terms with the role of language in thought."

Haldane added: "As I understand it, Turing's Imitation Game is based exclusively on a linguistic exchange between whatever or whoever is on the other side of the screen and the human interrogator. This certainly seems to suggest that any type of thinking machine would have to have human language capabilities as a minimal requirement to be considered intelligent. Yet Wittgenstein tells us that such capabilities can only arise from a shared form of life, one that seems to exclude machines. It appears there is a direct contradiction here to the notion of a machine thinking like a human. For myself, I would like to hear more discussion on this point to see if we can't find some way out of the dilemma."

Chapter Four

The Meat

Meaning and Machines

As Simmons bustled about the table, removing dirty silverware and bowls and setting out new knives and forks for the meat course, Wittgenstein excused himself to the lavatory while Turing wandered away from the table into the sitting room to gaze briefly out of the window at the dying storm. With the two antagonists temporarily away from the table, Snow took the opportunity to ask both Haldane and Schrödinger their views of the arguments put forth thus far.

"Turing seems to be offering a view of the brain as being something like a symbol-processing machine," said Snow, "in which strings of 0s and 1s on a long tape somehow come to actually mean things as different as a cup of tea or a journey to China. So I'd like to . . . "

"What confounds me," interrupted Haldane loudly, "is that these are literally meaningless symbols, Snow. Turing could equally well have chosen to use dots and crosses or stars and squares, or even colors like black and white as the markers on his tape. How can he believe that his machine will ever truly understand these symbols as having content? How can abstract strings of symbols actually *mean* something in and of themselves?"

As Wittgenstein and Turing returned to the room, Schrödinger laid the matter on the table for their consideration. "While you were away we have been puzzling over how this computing machine, which seems able only to rearrange patterns of 0s and 1s on a tape, can ever come to acquire an *understanding* of what these patterns actually represent. It's a simple matter to see how thinking beings like us standing outside the machine can attach meaning to these patterns. But how can a computing machine like Turing's ever come to this kind of understanding internally simply by moving 0s and 1s about on its tape? That is the question that seems to puzzle us."

Before Turing could reply to this puzzle, Simmons re-entered the room bearing a stack of clean plates, and proceeded to carve generous slices of thick, juicy roast beef and place them on the plates, accompanied by helpings of roast potatoes and green beans. Breaking into the lull in the conversation brought on by these culinary machinations, Wittgenstein began to attack the adequacy of Turing's test for machine intelligence.

"Turing has told us that all his machine requires to be regarded as 'intelligent' is that it be able to fool us into thinking it is human by giving responses that are indistinguishable from those we would expect to receive from a fellow human being. So if we can't reliably tell the difference between the machine's responses and a human's, then either the machine is intelligent or the human is not. Is that right, Turing?"

"Yes, that is a fair summary of my argument for the Imitation Game as a test of intelligence," Turing agreed.

"Well then," said Wittgenstein, "let's consider a different sort of game. In fact, it's one that uses exactly the same closed room and communication scheme of this Imitation Game." Pushing aside his plate, Wittgenstein outlined the following scenario as Snow signaled Simmons to fill everyone's glass with a hearty Burgundy, the perfect complement in his view to the juicy slices of rare roast beef sitting on their plates.

"Suppose we sit Snow down in Turing's closed room," continued Wittgenstein. "Inside this room is the teletype machine, together with a large book containing two columns of hieroglyphic symbols. Now let's take Schrödinger here, a man who understands such arcana, and ask him to type hieroglyphic expressions on the teletype keyboard. Perhaps Schrödinger will oblige us now by writing down such an expression on Turing's note pad."

Schrödinger quickly jotted down the following collection of hieroglyphic symbols:

Wittgenstein continued: "Seeing this statement on the teletype, Snow opens the book and looks for Schrödinger's string of symbols in the left-hand column. When he finds this string, he types out the string listed opposite it in the right-hand column. The corresponding set of symbols for this expression might be:

![hieroglyphic symbols]

"What we have here is a written exchange between Snow and Schrödinger of exactly the type Turing wants us to believe constitutes the core of intelligence, and that forms the basis of his Imitation Game. After several exchanges of this sort, Schrödinger has no reason to believe that there is anything other than a skilled Egyptologist at the other end of the teletype, since he immediately recognizes the replies from inside the closed room as being simply a sensible reply in hieroglyphics to the expression he has been typing in from outside the room. But, in fact, there is no Egyptologist at all inside the room. There is only Snow, furiously typing out one set of symbols that is totally meaningless to him in reply to another, equally meaningless, string of symbols that is presented to him on the teletype."

Looking up from his plate, Haldane asked if Wittgenstein was claiming that Snow's activities inside the 'Hieroglyphic Room' were the same as what the scanning head did on a Turing machine when it moved back and forth, reading, writing and erasing symbols on the tape.

"Exactly my point," replied Wittgenstein with an almost evangelical intensity. "There is absolutely no difference between what Snow does inside the room when he reads the symbols on the teletype, consults the dictionary and puts a response on to the tape, and what the Turing machine does when it reads a symbol on its tape, consults its program and then writes a new symbol on the tape."

Snow then asked: "So what's at issue here is that if I have no *understanding* of what these symbols mean, then the Turing machine can't understand the symbols on its tape either. Is that your assertion?"

"Quite," replied Wittgenstein. "And if there is no understanding, there can be no thinking. Neither Snow nor the machine is thinking, because neither of them is able to actually understand what the strings of symbols they are processing represent. So I ask you, where is the semantics, either in the room or in the machine? The answer is that the semantics are nowhere; no amount of syntactic shuffling of symbols can ever give rise to semantics. And since all that Turing and his machine are capable of doing is syntactic symbol manipulation, there can be no thinking by such a machine."

"Let me see if I've got the gist of this Hieroglyphic Room business," said Haldane. "Wittgenstein seems to be making two fundamental points here. The first is that from Schrödinger's perspective as an observer *outside* the room, whatever or whoever is inside the room passes the test for intelligent behavior specified by Turing's Imitation Game, because that entity gives sensible replies to queries posed to it from the outside, and these replies are indistinguishable from those we would expect to see from an intelligent human Egyptologist. But Wittgenstein's second point is that from Snow's perspective sitting *inside* the room, there is no understanding at all; hence, there is no thought because no amount of syntactic shuffling about of symbols will ever enable Snow to know what the string of symbols actually means."

"I don't think you can just blithely ignore the enormous difficulties here that are associated with the 'simple' act of looking up all the various matching symbols in the hieroglyphic dictionary," said Snow.

"Indeed," added Haldane. "Even if you could get around the physical constraints of the size such a dictionary would have to be, there is an added problem with Turing's 'outside-the-system' test for intelligence."

"And what might that be?" enquired Turing, rather peevishly.

"Well," continued Haldane. "Suppose you mapped out in some kind of tree structure every possible conversation of, say, one hour in duration. Then by following this tree structure, the machine could interact with the interrogator in a way indistinguishable from the way an intelligent human would do so. Yet the machine would be simply making its way, branch by branch, through this tree. This suggests to me that the machine has no mental states at all."

"So you think that what's wrong with the tree structure, and, hence, Turing's test, is not the behavior it produces but the *way* it produces it. Is that right?" asked Snow.

"Precisely," replied Haldane. "In my view, to call a behavior intelligent is to make a statement about how that behavior is produced."

Schrödinger then put the cat among the pigeons when he stated: "To get back to Wittgenstein's first-person argument against the machine having cognitive states, let me say that for myself I find this line of argument similar to claiming that by chopping off the legs of a fly, I've made the fly go deaf. And why not? After all, it doesn't seem to jump any more when I clap my hands."

"What do you mean?" asked Snow.

"Well, Wittgenstein's argument does seem to have a sort of superficial validity. But digging beneath the surface, I think you'll find it rests on very shaky logical grounds," replied Schrödinger.

Before Schrödinger could go on, Turing, who had remained silent during these deliberations on his proposal for identifying intelligence, put down his fork with a clatter, shoved aside his plate and rejoined the discussion.

"Wittgenstein's position would be clearer to me if I were to phrase his line of reasoning in purely axiomatic terms." He reached across the table for his note pad, and proceeded to quickly outline Wittgenstein's three major assumptions and the logical conclusion that followed from them:

The Hieroglyphic Room Argument

Axiom 1: Programs are purely syntactic objects.

Axiom 2: Human minds have semantic content.

Axiom 3: No amount of syntax can generate semantics.

Conclusion: Programs are neither necessary nor sufficient for minds.

"Seductive as this argument appears," Turing went on, "I agree with Schrödinger. Wittgenstein has managed to sneak some hidden assumptions into the axioms that cast serious doubt on the conclusion."

"For instance?" barked Wittgenstein, in his characteristically pugnacious tone.

"In my view, your argument draws attention to the wrong system, Wittgenstein. You speak about Snow being inside the Hieroglyphic Room, receiving and sending strings of arcane symbols on the teletype. You then argue that since *his* mental states inside the room clearly have no understanding of hieroglyphics, then there are no computational states anywhere in the situation that are semantically tied to the exchange of information between Schrödinger outside the room and Snow inside it."

"That is certainly my claim," Wittgenstein replied, ominously waving his knife and fork in Turing's general direction. "There is no meaning anywhere in this situation."

"I beg to disagree," argued Turing. "There *are* such computational states; they are the states of the entire room. It just will not do to direct your argument to the states of Snow's brain alone, since he is only a *part*

of the Hieroglyphic Room. This would be like center-
ing attention on just the scanning head of a computing machine, ignoring completely the tape. But if you
consider the room itself as a complete system, then this
system does possess states with semantic content."

Glaring across the table at Turing, Wittgenstein
asked "And just what are those states?"

"Allow me to explain."

Turing then went on to say that just as the computing machine consisted of the scanning head, the
program and the tape, the room consisted of the teletype machine, the dictionary and Snow. So one had to
consider the states of the *entire* system in both cases,
not simply those of a part of the system.

Haldane then broke in to this explanation, asking
Turing: "You're saying that the relevant system is Snow
plus the book plus the teletype plus the wall between
all this and Schrödinger. Is that it?"

"Sheer nonsense," objected Wittgenstein. "The
physical paraphernalia has nothing to do with my point.
My argument doesn't change in the slightest if we elim-
inate the room by having Snow simply memorize the
dictionary of hieroglyphic responses to whatever inputs
happen to be presented."

Snow thought to himself about how he could possibly do this without accidentally learning hieroglyphics in the process. After all, he thought, I have only
one set of perceptual and motor systems. So I should
eventually make the appropriate associations between
the symbols coming to me through my sensory organs
and the 'correct' symbols that I send back. Almost as if
he were reading this line of reasoning taking place in
Snow's mind, Wittgenstein answered this objection.

"And if you think that Snow would somehow have
to actually learn hieroglyphics in the process of mem-

orizing the dictionary, he wouldn't. For example, we could imagine that he develops another, completely separate, cognitive system. This system has the effect of producing another person inside his body, a person who is inaccessible to the Snow we have sitting with us here at the table. This leads to a multiplicity of systems sharing a single physical body—each with its own semantics."

During this interchange, Schrödinger absently polished his spectacles with his napkin while he pondered the coherency of Wittgenstein's thought experiment, trying to pick apart the various threads in the argument for their logical consistency. It seemed to him that the overall chain of reasoning didn't quite hang together—but he wasn't entirely sure *exactly* where the flaw was either. So while the others were occupied in debating whether or not the Hieroglyphic Room as a single entity truly understood hieroglyphics, Schrödinger put together an analogous argument in physics that he hoped would clarify the logic that underlay the Hieroglyphic Room.

"Suppose," said Schrödinger, "that instead of considering the elusive property of meaning we look at a physical property like luminance. Let's try to mimic Wittgenstein's arguments in the context of the physics of light and see where it leads."

Borrowing Turing's note pad, he continued: "Let's consider Turing's axiomatic formulation of the Hieroglyphic Room and transfer it to what we might call the Luminous Room. We then end up with the chain of reasoning I've sketched on this page. You'll have no difficulty seeing the correspondence between this argument and the one put forth a moment ago by Wittgenstein with his Hieroglyphic Room."

The Luminous Room Argument

Axiom 1:   Electricity and magnetism are forces.

Axiom 2:   The essential property of light is luminance.

Axiom 3:   Forces by themselves are neither constitutive of nor sufficient for light.

Conclusion: Electricity and magnetism are neither constitutive of nor sufficient for light.

"If Wittgenstein had raised this argument in the last century, shortly after Clerk Maxwell's suggestion that light and electromagnetic waves are one and the same thing, the Luminous Room might well have served as a seemingly airtight objection to Maxwell's claim. But he would have been wrong," concluded Schrödinger with some authority.

"Yes, I see what you mean," said Snow as understanding slowly dawned on him. "If a man in a dark room moves a magnet up and down, Maxwell's theory claims that this creates a spreading circle of electromagnetic waves. So the room will be luminous. But from playing with magnets we all 'know' that waving a magnet around in front of us produces no luminance at all. So it would appear inconceivable that one could create *real* luminance by just moving magnetic forces around. But, in fact, that is exactly what happens."

So here was the crux of Schrödinger's thought experiment. How would Maxwell have replied to this challenge to his theory of electromagnetism and light as being *exactly* the same phenomenon?

Turing immediately jumped to the defense of Maxwell's theory, asserting that "One way Maxwell could counter Schrödinger's argument would be to insist that the Luminous Room experiment doesn't properly illustrate the physical phenomenon of luminance because the frequency of oscillation of the magnet is much too low. Consequently, it would generate waves of energy whose rate of oscillation is too slow to be perceived by the human visual system."

Schrödinger fended off this line of attack, noting that "How fast the magnet moves has nothing to do with things. By Maxwell's theory, the room with the moving magnet contains everything essential to light. So you can't defeat the Luminous Room just by saying the magnet is moving too slowly."

"But the biology of our particular kind of human nervous system should enter into consideration here somewhere," objected Turing. "For instance, Maxwell could claim that the room really is illuminated, but at a frequency of radiation that's too low and at a level of intensity too weak for the human visual system to detect. Of course, in Maxwell's time—in the 1860s—such a reply would probably have brought on howls of laughter and jokes to the effect that the room is pitch black. But, of course, we all know now that Maxwell was absolutely correct. And that is the point of this Luminous Room argument. Let's consider for a moment what this thought experiment really tells us.

"First of all," Turing continued, "despite its intuitive plausibility, Axiom 3 of the argument about the inadequacy of forces to create light is totally false. Secondly, the Luminous Room tells us absolutely nothing about the nature of light. Finally, what's needed to settle the matter of whether or not the room is illuminated is a research program addressing the conditions under which the behavior of electromagnetic waves does indeed create luminance. So even though Wittgenstein's Hieroglyphic Room appears to be 'semantically dark', it's just not possible to justify on the basis of this appearance the claim that symbol manipulation can never give rise to real meaning."

Having just polished off a yeoman's helping of meat and potatoes, Haldane was well fueled and ready to return to the discussion. "It seems to me," he said, "that all this talk about stringing symbols together and whether or not these strings of symbols mean something misses the point of human thought."

"In what way?" asked Snow.

"To my eye, thinking involves a lot more than merely computing the right functions. To think requires the ability to perceive the world around you and to move about in that world. The problem with Turing's computer is that it just sits there."

"Are you saying that if we put Turing's machine inside a 'mechanical man' having artificial sensory apparatus, a kind of 'robot', you might call it, we would then have something that thinks in a way that's fundamentally different from a disembodied brain?" asked Schrödinger, with some degree of incredulity.

"Perhaps," replied Haldane. "Such a robot would be a lot closer to my idea of a thinking object than a machine that just sits there and cogitates by following a set of rules."

Turing jumped in to support Haldane, introducing a new twist to his argument in favor of a thinking machine. "A great positive reason for believing in the possibility of making machinery that thinks is, as I said earlier, the fact that it's possible to make machinery to imitate any small part of a man."

"Like using a microphone to imitate an ear or a camera to perform the same function as the eye," said Snow, recalling Turing's earlier comparisons.

"Precisely," replied Turing, warming to his topic. "For thinking, we are chiefly interested in the nervous system. And there seems no special obstacle to building a complete electronic simulation of the entire nervous system. So one way of setting about the task of building a thinking machine would be to take a man as a whole and to try to replace all of his parts by machinery, one by one."

"This would be an enormous undertaking," noted Schrödinger with some suspicion. "You would have to include television cameras, microphones, loudspeakers, wheels and all types of servomechanisms to control these devices, not to mention an electronic 'brain' to coordinate things."

Wittgenstein cast a somewhat baleful glance at Turing, asking, "Are you seriously suggesting that by turning such a device loose in the countryside, like unleashing Frankenstein's monster, that it would be able to learn things for itself and thus become 'intelligent'? Allow me to re-emphasize the crucial point that such a literally incredible machine would still have no contact with many things of interest to human beings. It's absurd to imagine that whatever kind of 'intelligence' such a mechanical contraption achieved, it would in any way be similar to that possessed by a human."

"I'll admit this point," Turing said. "I am proposing to see what can be done with a 'brain' that is more or less without a body, providing it with, at most, organs of sight, speech and hearing. Of course, with such restrictions on sensory input we must find suitable branches of thought for the machine to exercise its powers in. The most appropriate seem to be games like chess and checkers, as well as other basically linguistic tasks like language translation, cryptography and mathematics. Earlier we discussed the work of McCulloch and Pitts on the creation of artificial neural networks that would mimic the brain's circuitry in electronic components rather than 'wet' neurons. This is the type of electronic brain I'm thinking of."

Attempting to stanch the flow of Turing's stream of consciousness about mechanical thought, Snow began refilling everyone's water glass while he struggled to clarify the general connections between sensory organs and thought. "If I understand Haldane's point, he is saying that in order to perceive something like this pitcher of water, it's necessary to do more than simply execute a function; you must *interact* somehow with the pitcher. For example, your visual system—lens, retina, optic nerve and so on—has to process the light reflected from the pitcher, your motor system must manipulate it to fill your water glass, and so on. If a machine could do this, then it would necessarily know what these visual signals *mean*. Is that the thrust of your argument, Haldane?"

"Basically, that is the essence of it," replied Haldane.

"Well, I flatly disagree," said Turing with a note of finality in his voice. "First of all, it's not true that a computer just sits there. If it did, we couldn't interact with it by giving it programs, inputting data and getting output. So you cannot claim that computers do not think and still maintain that computation plus interaction is enough for thinking, since we already have interaction with computing machines."

Schrödinger then asked: "So in order to justify his belief that a robot could think but a machine cannot, Haldane would have to show why these standard kinds of interactions with the machine are not of the 'right sort'. Moreover, he would have to show what the right types of interaction are."

"My point exactly," replied Turing.

"Well, Haldane, what *do* you think constitute the right types of interactions?" asked Snow.

Haldane responded: "Let's suppose that the sensory inputs received by the human eye are processed by the visual system as *analog* signals rather than digital. Then the signals would be transmitted to the brain as infinite-precision real numbers rather than the finite-precision integers which is all that a digital processor like Turing's machine can produce. In that case, the brain would be making essential use of what amounts to uncomputable quantities, quantities that could never be obtained by following a set of rules on a digital machine."

"Are you suggesting that the only way the visual system can recover the right information from the environment is to employ a kind of infinite-precision processing? If so, let me remind you of the physics of this situation" said Schrödinger with some authority. "Even so-called analog signals of the type you're suggesting are subject not only to noise from their environment, but also to measurement errors that limit the precision to which they can be measured. So it's just not the case that you could transmit infinite-precision real numbers by such devices. What you're talking about is a mathematical idealization, a physical fiction. Do you ... "

"Leaving this physical constraint aside for just a moment, are you suggesting that there is a crucial component of thinking that is essentially uncomputable by a machine like Turing's?" enquired Wittgenstein, breaking in to Schrödinger's objection with one of his own.

Before Haldane could reply to these queries, Snow began furiously tapping on his water glass to attract the attention of the combatants in this increasingly heated and speculative discussion, which seemed to him to be veering off into the esoteric. He felt it was time to try to summarize the ideas and conflicting views that had been buzzing about the table for the past half-hour or so and bring the deliberations back down to earth.

"Let me see if I can bring together the claims and counterclaims that have been circulating here this evening. Turing began by offering his Imitation Game as a way of identifying intelligent behavior—in men and machines. This is a third-person type of test, sitting right at the heart of the behaviorist tradition in psychology. As I'm sure you are all aware, this tradition focuses attention on an object's external behavior in response to sensory stimuli. Turing's argument is that if extended interrogation of the machine does not allow us to distinguish the machine's replies from those given by a human, then we must declare the machine to be 'thinking'. Is that a fair summary of the position you've staked out, Turing?"

"Quite satisfactory," Turing responded.

"All right," continued Snow. "Wittgenstein then produced a first-person counterargument based on his rather fanciful, but extremely instructive, Hieroglyphic Room. He asks us to imagine the workings of a computing machine *from the inside,* as it moves symbols about from place to place on its tape in accordance with the rules coded into the machine's program. According to this picture, the machine cannot possibly have any understanding of what the symbols actually *mean;* so, it cannot possibly be thinking. Is that the crux of your claim, Wittgenstein?"

"A crude caricature of my position, barely sufficient for this discussion," replied the Austrian philosopher, chafing somewhat at Snow's oversimplification of his thought experiment. "But I'll accept it for the sake of argument."

Passing over these mutterings in silence, Snow continued: "Now we come to the objections raised against the Hieroglyphic Room, which, on the principle that the enemy of my enemy is my friend, I think we can regard as tantamount to being arguments in favor of Turing's Imitation Game. First off, there is Turing's claim that while my brain inside the room by itself may have no computational states with semantic content, the entire room, consisting of the walls, the teletype link, my brain and the dictionary of transforms, certainly does possess such states. We might term this the 'systems' reply to Wittgenstein, since it asserts that the complete system of my brain plus the book of transforms plus all the rest of the room constitutes an object with semantically laden computational states.

"Next, Schrödinger conjured up what he calls the Luminous Room, in order to draw an analogy between luminance, a physical property of electromagnetic radiation, and Wittgenstein's mental property of a brain, namely meaning. If I understand Schrödinger's reasoning, the conclusion is that it's difficult for us to believe that the forces of electricity and magnetism really are the same thing as light. So, by analogy, we find it equally difficult to accept that meaningful content can arise simply from the interaction of raw symbols. But according to the logic of Schrödinger's Luminous Room, it does."

Wincing slightly at this last part of Snow's summary, Schrödinger clarified his position. "I'd prefer to say that I see no logical obstacle to meaning arising from a machine that simply processes sequences of symbols into new sequences formed from the same symbols. As to whether meaning actually *does* arise out of such operations, well, that's an empirical question. It can only be settled by observation and experiment."

Nodding a silent thanks to Schrödinger for helping to straighten out his argument, Snow glanced over at the doorway and said, "Before I go on with this summary, I see Simmons is eagerly waiting to take away our plates and get on with the next course. So would any of you care for another helping of this delicious roast beef before we commend it to his custody?"

"Absolutely first-rate beef, Snow," remarked Haldane. "It's a pity we don't get it on a more regular basis. But I've already had more than my share for one sitting." As the others murmured their agreement, Snow motioned for Simmons to remove the plates and continued his summary of the arguments in favor of and against both the Imitation Game and the Hieroglyphic Room.

During Snow's recitation, Turing was squirming and fidgeting in his chair like a man crawling with fleas. He finally broke in to the middle of Snow's statement, blurting out: "I strongly object to Wittgenstein's unfounded claim that the third-person, external view of a thinking machine represented by my Imitation Game doesn't capture the essence of how we distinguish a thinking being from one that's not. In contrast to Wittgenstein, I believe that this external view is the *only* valid one."

"Perhaps you'd care to enlighten us as to why you continue to cling to this fantastic idea," responded Wittgenstein in an eerily calm tone.

"All right," said Turing. "Consider a grandfather clock, like the one sitting over there against the wall. Viewed from the outside, this clock tells us the time. I think we can all agree that that is its primary purpose."

"Obviously, clocks tell time. That's what they're for," said Haldane. "But what does timekeeping have to do with the Imitation Game?"

"Looked at from the inside, the clock no longer performs that function," said Turing. "If I take the clock apart and lay all the components on this table, it ceases to be a timekeeper. Its ability to keep time depends on the parts being assembled and interacting correctly with their neighbors. So the clock's ability to tell time— and for us to recognize this—depends on our being on the *outside* of the set of interacting pieces. In this sense, timekeeping is an external, holistic function of a clock; it cannot be recognized by standing inside the collection of gears, pulleys, springs and the like. It is an *emergent* property of these pieces and their interaction with each other."

Snow took back the floor, saying: "Are you claiming that situation relates to the Hieroglyphic Room? That the room's competence in hieroglyphics is seen when we view it from the outside as a complete whole. Yet if it is 'taken apart' it no longer has this ability? Is it your contention that comprehension of hieroglyphics is a holistic function of the room?"

"That is precisely my position. I'm sure that if you chop up a brain surgically, it wouldn't display much capability of understanding anything," responded Turing. "And neither does this Hieroglyphic Room, if you look only at its components separately."

Moving from the table to the window, Schrödinger, appeared to be lost in thought as he stared out at the storm, which seemed to be picking up again following a brief lull. After some moments of silent reflection, he returned to the table and presented the group with a new slant on the inside/outside argument between Turing and Wittgenstein. "Hindus and Buddhists have a belief that all is ultimately form," he said quietly. "They argue that there is no such thing as content or meaning. What we perceive as content is merely the external form of another layer or level. They liken this to an onion. Peel away one skin and there is another underneath.

And if you peel away *all* the skins, there is absolutely nothing inside. So as I said a moment ago, I see no logical obstacle to meaning arising from symbol processing, for the very simple reason that there may well be no such thing as meaning."

Again sensing the argument taking a rather pronounced turn to the philosophical, Snow tried bringing things back to material and practical reality.

"All these arguments have assumed that the computing machine has only a certain stylized way of interacting with its environment. Basically, the environment, consisting of the machine's operator, places symbols on the tape. The machine then communicates with the operator by writing some other symbols on its tape. Haldane raised the question of whether this very circumscribed sort of interaction is too impoverished for thinking to occur. He told us that perhaps some type of robot having sensory apparatus like eyes and ears might be able to think, but not an object like a machine that just sits there, a passive lump of glass and metal and ceramic. In other words, sensory interaction with the environment, human-style, is a necessary condition for an object to be able to think."

"Just so," chimed in Haldane. "Thinking is a combination of computation *and* interaction. Sensory inputs to the brain *do* matter."

Turing couldn't resist adding, "Perhaps Haldane's right about this. But what he's not saying is that if these stylized interactions the machine uses to make contact with the world are not the 'right sort' for intelligence, then what are? Must we *duplicate* human sensory apparatus? Is it necessary to give the machine a sense of taste, touch and smell? Or is it enough for it just to be able to see and hear? And if we do have to create a machine version of these five senses somehow, why should we think that any of these human senses are not themselves computational processes?"

Schrödinger then put forth the idea that "Perhaps sensory organs like the eyes and ears have to transform information about the world into a special form that the brain can use. If so, I suppose it's possible that this transformation process could transcend the kind of computation that can be done by Turing's machine."

"If that is the case," said Snow, "then we would be in a situation in which thinking requires sensory inputs, which in turn cannot be obtained as the result of following rules, that is, of any sort of computational process. This would certainly exclude the very idea of a thinking machine. But it's asking a lot to accept either of these hypotheses, let alone both."

"How odd that what's crucial to cognition should turn out to be *exactly* the information we cannot, as scientists, measure with our instruments," remarked Schrödinger drily. "While I concede that this *might* turn out to be the case, there is certainly no conclusive evidence at present—in fact, *any* evidence—that unmeasurable quantities are an essential ingredient of human cognition."

Haldane closed down this particular line of attack on the possibility of a thinking machine, noting that "One of the cherished principles in science and philosophy is Ockham's Razor, to the effect that an explanation of anything should be as simple as possible—but no simpler. Explaining cognition in terms of uncomputable sensory inputs strikes me as an excellent example of how to violate this principle. Until I see something on this point that looks like evidence rather than mere personal opinion and unbridled speculation, I'll continue to believe that everything is computable until proven otherwise."

Wittgenstein had been brooding silently during most of this discussion about sensory apparatus and thinking. Suddenly he burst out of his lethargy, telling the group that "Every human thought is intimately connected to its linguistic expression. There can be no thought without language. I've sat for nearly two hours now listening to nonsensical chatter about machines, sensory organs, symbols and the like without hearing a single word about language. How can anyone speak of a 'thinking machine' without considering the language by which its thoughts are represented? Everything said around this table so far is complete nonsense without coming to an understanding of this point."

A bit taken aback by the abruptness of Wittgenstein's outburst, Snow tried to bring the discussion back to this point.

"If I'm not mistaken," he said, "it was Aristotle who said that human beings are essentially language-using animals. If this is indeed the defining feature of what it means to be human, then it seems to follow logically that for a machine to duplicate human thought it must have the 'gift' of language—just as Wittgenstein says."

"But what *kind* of language would such a machine use?" enquired Haldane, sharpening the thrust of Wittgenstein's objection. "And is that type of language compatible with the language used by the human brain? It seems to me that those are the issues that need clarification if we're to understand the potential of Turing's machine for human thought."

"Indeed, that does appear to be the question," said Snow. "And this appears to be an excellent time to take a short break from our deliberations before Simmons serves the salad. Perhaps we can all consider this issue of how language enters into human thought processes, and share those views when we come back to the table.

But for now, I suggest we refill our glasses, stretch our legs a bit, and return to the table in, say, ten minutes or so to continue pursuit of this crucial point."

Chapter Five

The Salad

Language and Thought

Returning to the table, at each place the guests found plates of leafy green salad mixed with tomatoes and lightly sprinkled with oil and vinegar. As they passed around the salt and pepper grinders and began eating this hard-to-come-by delicacy, Snow re-opened the theme of humans and language.

"Wittgenstein has reminded us that what distinguishes humans from other living things is our ability to use language to express our thoughts and to communicate them to others of our kind. So if Turing's machine is going to pass the test he outlined and fool us into thinking it's human, then it seems to follow that it will have to have human-like language capabilities. Can we all agree on this?"

"Just so," mumbled Haldane through a mouthful of salad. Pointing his fork at Wittgenstein as if he were sighting a rifle, Haldane continued: "I think Wittgenstein has hit the nail right on the head. How can it make sense of any kind to regard a machine as thinking like a human unless we can communicate with it in *human* language, not the bizarre programming language of long strings of 0s and 1s that Turing offered earlier? Give me warm, idiosyncratic words and sentences any time, not sterile, remote, pristine strings of 0s and 1s."

Tapping the side of his plate with a knife to gain the group's attention, Schrödinger interrupted the discussion for a point of clarification. "Before we go into the matter of human language and how Turing's machine might acquire it, could we possibly clear up the distinction between the way humans use language for communication and the way other species, like birds or ants, communicate. There are those who use the term 'language' to describe these forms of animal communication as well. So I'd like to make sure we're all talking about the same thing when we speak about 'human language' here, as opposed to these other forms of communication. Just what is it *exactly* that separates human forms of communication from these other types?"

"Excellent point, Schrödinger," noted Snow. "Haldane, would you care to elaborate on this distinction?"

Why me? thought Haldane. I'm no specialist on languages. Never one to decline the podium when it was offered, though, he turned away from his salad and rose to Snow's challenge.

"Ours is the only species able to use language in the full sense, meaning that human beings can employ a connected set of conventional signs to communicate. While it may be that other animals such as birds and monkeys communicate by signs, like birds that shriek when something dangerous is present, or bees that communicate the location of food by performing an elaborate dance, these signs are not really language.

"And why not?" asked Turing.

"The reason rests with the terms *connected* and *conventional*. When we speak of a 'conventional' sign, we mean the sign has no natural connection with the thing being talked about. So, for instance, the word 'water' in no obvious way relates to the liquid material to which it refers; the word has no intrinsic meaning, and we could just as well employ the French term *eau* or the German *Wasser* to describe what we mean. This differs considerably from the kind of sign that one might make to indicate 'water', such as making a wavy motion with the hand. The use of the symbol 'water' also differs from screeching because it is a learned and agreed-upon code for a particular liquid rather than being a spontaneous reaction."

Continuing his discourse, Haldane noted that the term 'connected' means that humans use language as a complex arrangement of signs that can be employed in an unlimited number of combinations with each other. It is this connectivity that allows humans to form combinations to express virtually any thought that a human brain can have, ranging from a charging bull to a peaceful summer evening to the smell of freshly mown hay. The unbounded possibilities for making distinctions and forming combinations of these distinctions by using grammatical rules and structures is then the second feature setting human languages apart from more primitive types of communication systems.

"So," concluded Haldane, "while some may argue that the properties characteristic of human language—conventional signs and connectivity—differ from animal communication only in degree, they are nonetheless real and are what give human language its expressive power. And I think you'll see ... "

Interrupting this discourse, Wittgenstein said quietly, "A dog cannot lie. But neither can he be sincere."

Taken aback by this strange proclamation, the rest of the group silently waited for Wittgenstein to complete whatever thought had compelled him to make such an unprovoked, out-of-the-blue statement. Staring down at the table, after a long pause he continued.

"A dog may be expecting his master to come. Why can't he be expecting him to come next Wednesday? Is it because he doesn't have language? If a lion could speak, we would not be able to understand it. Why do I say such a thing, Haldane? Why do I say it?"

"Damned if I know, Wittgenstein. But if I could understand you I shouldn't think I'd have much trouble with a lion."

With a fierce glare Wittgenstein silenced the beginnings of a faint chuckle from Schrödinger at this show of asperity on Haldane's part, and forged onward: "To have a language is to have a way of life. Everything we say is totally bound up with what we do. How can I know what world a lion inhabits? And so how could I possibly hope to understand its language? Do I fail to understand it because I can't peer into its mind? Because there is something behind its words that I cannot grasp?"

"Perhaps it would be best if we return to this point a bit later on," said Snow quickly, attempting to head off an extended monologue by Wittgenstein on the nature of language. "At the moment, I see our concerns as being with the relationship between language and thought. In particular, how the brain links language to thought. I'm sure Turing has ideas on this matter."

"Indeed, I certainly would like to say something about this," Turing replied. "As I stated earlier, I don't see any difference that matters between the workings of the brain and those of a computing machine—including the way the brain uses language to communicate thought." Picking up the water pitcher sitting in front of him, Turing went on. "Basically, my view is that a concept like *water pitcher* is coded in the brain by a particular collection of neurons being ON and OFF. This pattern then interacts with other neuronal patterns, for example, the pattern for *glass* and the pattern for *pour,* to create more complex thoughts like *pour water from the pitcher into the glass.* I believe a computer could think in the same way—by manipulating various patterns in its memory, making them interact with each other, assembling and disassembling what in a human brain we call 'thoughts'."

Schrödinger then asked: "So you're saying there is a kind of 'language of thought' in the brain. Patterns of neurons code for all the various concepts of the world, and the brain assembles these patterns in different ways according to some rules—a 'grammar of thought', you might say—to give rise to what we regard as 'thinking'. Is that your view?"

"Yes, I believe that's a good account of what I have in mind," replied Turing.

Stabbing at a piece of lettuce with his fork, Wittgenstein recalled somewhat sourly that this view of thought could be regarded as a kind of 'mentalese' coded into the brain. In this sense, Turing's view on language and thought was reminiscent of Wittgenstein's own picture theory of language put forth in his book *Tractatus Logico-Philosophicus.* As he went over all the reasons for why he later rejected this view of language, Snow, almost telepathically, voiced some of the very same concerns.

"What Turing has just suggested seems very close in spirit to the vision of language that Wittgenstein proposed some years ago. By my recollection—and perhaps Wittgenstein can correct me if I'm wrong—this 'picture theory' of language asserts that language and reality have a common logical form. This then implies that language mirrors the world, and that linguistic propositions picture facts."

Haldane then interrupted: "Do you mean the point at which language hooks on to the world is through the relationship between an object of the world and the name the language attaches to that object?"

"Precisely," replied Snow. "The way reality is projected into propositions is for the real world and the language to have a common logical structure. Thus, linguistic statements are meaningful when they can be correlated with the world. So, for instance, I can meaningfully say, 'The Royal Albert Hall is in London.' But it is meaningless to say, 'Is Royal the Hall Albert London in.' Of course, different grammars, or rules, could be concocted, within which this last statement would be meaningful. But within the conventional grammar of the English language, it has no logical structure at all."

"So the main claim of this picture theory of language is that there must be something in common between the logical structure of the language and the logical structure of the fact that the language asserts. Is that it?" asked Schrödinger.

"That's exactly it," interrupted Wittgenstein. "And that relationship between the fact and its expression in language is *precisely* what can never be expressed in language. Words of a language can never express this correspondence. So give up this vision of language mirroring logic; it's complete nonsense."

"Wait just a minute," said Snow. "Let's follow up this theory in relation to Turing's idea of thought as being akin to pushing around and combining various symbolic representations of real-world objects in the brain. This fits rather well with your picture theory.

All we need do is associate Turing's symbolic coding of objects with the pictures of your theory."

Throwing down his napkin in frustration, Wittgenstein exploded: "You have completely misunderstood what I mean by a 'picture'. The picture is *not* the representation of the object in the mind; I am not talking here about pictorial images of trains, tables or top hats. I am referring to the impossibility of ever expressing the relationship between the object and the name the language attaches to it. This relationship can only be *shown*, never expressly stated in language. It is that 'showing' rather than 'speaking' that *is* the picture. But this is an erroneous picture of language and I completely renounce it."

"No pun intended, eh, Wittgenstein?" joked Haldane, at this unwitting dual reference by Wittgenstein to 'pictures'.

"Please excuse my confusion regarding the role of 'pictures' in the theory of language you presented in the *Tractatus*," apologized Snow. "But since you now seem to believe that this is a wrong-headed theory of the relationship between language and thought anyway, perhaps this is a good time to return to your earlier statement that a language is intimately tied up with a form of life. Presumably this relates to your current position on the role of language in thought. If you would be kind enough to explain yourself on this matter, I'm sure we would all greatly benefit."

As if unleashing a chained dog, at Snow's request Wittgenstein nearly jumped out of his chair in eagerness to take the floor. Leaning forward and addressing the group with great intensity, he said: "Language isn't a picture at all. Rather, it is a tool, a precision instrument."

"But an instrument for what?" enquired Turing.

"For making judgments. A creature without language, like the dog or the lion that I spoke of a moment ago, is, strictly speaking, incapable of being either right or wrong about something as being the case."

"Are you serious?" piped up Haldane. "Do you really believe that when my dog grabs at a branch on the street, thinking it's a bone, that he's not making a mistake? That he's not just plain wrong about the branch being a bone?"

"Well, certainly your dog can make a mistake. But what I'm saying is that it is not a mistake about the branch being a bone. To make a mistake about what is actually the case in this situation would be for your dog to apply the wrong concept. And your dog cannot possess the concept of a *bone,* even though he may be able to recognize one in a given situation."

Schrödinger then asked for further elaboration of this basic point. "Do you mean that the dog knows *how* to recognize a bone, but it doesn't know that this or that thing actually *is* a bone?"

"Exactly," responded Wittgenstein. "This type of knowledge is what language provides. It gives the *concept* of a 'bone'; that's why making judgments depends on the use of language."

Haldane then struck to the heart of the matter: "So if my dog cannot be mistaken about a branch being a bone, then it cannot have a language, at least not in the sense that humans have language. That's clear enough. But it raises the question of exactly what sort of organisms *can* possess this human-like sort of language. Is it simply a property of a particular kind of brain structure? Or is there something else involved?"

Now we're getting somewhere, thought Snow, as he helped himself to the pepper grinder and sprinkled some pepper on to his salad. Turing's whole argument about duplication of thought processes in a machine hinges crucially on this point, he mused. If thought requires language and language requires a particular type of brain structure, along with possibly other things as well, then maybe we can close out the matter of thinking machines by convincing ourselves that machines just don't have the right 'stuff' for language. I wonder if this kind of argument is what's behind Wittgenstein's obvious hostility to Turing's belief in the possibility of an electronic brain.

Just as Schrödinger cleared his throat and started to speak, Wittgenstein leaned over to Haldane and said, "The structure of the human brain is not the main point in language. Your dog's brain and mine probably don't differ all that much in structure, other than that mine is a bit larger. What's essential for language—and what your dog does not have—is the company of other language users."

"So, you're claiming that at bottom there can be no private language," said Turing. "Language is a matter of social convention. Is that your belief, Wittgenstein?"

"I'm saying that in order to make judgments, to agree on what cases are instances of concepts like *bone* and what cases are not, you need a language. To make these distinctions is what constitutes following a rule, as opposed to acting in an instinctive way like Haldane's dog. In genuine rule-following, there must be a distinction between *actually* following the rule, and what merely *appears* to be following the rule."

"You've lost me there, I'm afraid," said Haldane. "Why can't I have a private rule that I follow, a rule that's mine alone? And if I do have such a rule, why can't I tell whether I'm following it correctly or not?"

"A private rule is no rule at all. To engage in genuine rule-governed behavior, one must be a member of a community that upholds the rule and that serves as the authority as to what constitutes following and not following the rule," replied Wittgenstein.

"This is certainly a pivotal point in our discussion about language," said Snow. "But it still seems rather unclear to me how you can reject the notion that I might be following a private rule, unknown to anyone else here at the table."

"Yes," said Schrödinger. "Suppose we teach someone a rule by showing him the sequence of numbers 1, 2, 4, 8, 16, 32 ..., and then ask him to continue the sequence in 'the same way'. He will probably go on with 64, 128, 256 ..., seeing that each number in the original sequence was formed by doubling the previous number. But suppose that instead he wrote down the continuation 35, 38, 41, 44, ... If we point out to him that he is not continuing in the same way, he might say that he wasn't using the doubling rule, but rather the different rule: 'double the number up to 32, and after that add 3'. What I don't quite understand is why he can't have this private rule, and then simply tell us what he understands the rule to be."

Supporting Schrödinger's example, Snow said, "Yes, why can't he have this private rule and then reach a verbal agreement with us about which rule he is using on any given occasion?"

Wittgenstein's reply went straight to the heart of the reason why private languages are impossible. Looking around the table in his characteristically penetrating way, he answered Schrödinger and Snow.

"The reason he can't simply tell Schrödinger what rule he is following is that the verbal behavior involved in telling us which rule he is following is itself a rule-governed activity. So if we are uncertain about whether two people are following the same mathematical rule, consistency forces us to be skeptical about whether they are using the same linguistic rules when they discuss which rule is being meant—the 'doubling' rule or the 'doubling-to-32-and-then-adding-3' rule."

Puzzled by all this talk of rules, Turing asked, "Well, then how can we ever be certain that any of us are following the same rule as someone else? Or, even worse, how can I be sure what rule I myself am following?"

"There can be only one answer to this, Turing," replied Wittgenstein, in his best teacher-to-pupil voice. "For a number of people to be following the same rule is not for them to have the same private, inner conception of a rule. Rather, it means that they agree *in practice;* there is a public check that determines whether what they are doing is following the same rule or not."

So, thought Haldane to himself, if Wittgenstein is correct there can be no such thing as solipsism; there must be thinking beings other than myself. If making judgments requires a language, and if language is a rule-governed activity requiring a public check on its rules, then the absence of other people would mean that I could not be truly making judgments. So if I *am* making judgments—thinking—then there must be other intelligences. I can't even wriggle out of this conclusion by doubting that I'm thinking, for to doubt solipsism is, itself, a judgment. What an argument!

Vocalizing this line of reasoning to the group, Haldane declared: "If Wittgenstein's argument is correct, then rule-following in one person necessitates rule-following by at least one other person; there can be no private language or private interpretation of what it means to be following a rule. But this, then, seems to imply that a computing machine like Turing's cannot really *know* it's following a rule—or a program—and thus can never really be thinking like a human."

"Well," said Turing, with a puzzled look on his face, "regardless of whether language is a picturing relationship between words and the objects of the world, or is a kind of rule-based social consensus, as Wittgenstein now claims, I'd still like to hear Wittgenstein's view on whether thinking is done in *any* kind of language."

"The idea of a language of thought encounters a dilemma of its own," Wittgenstein replied. "On the one hand, while my words can be interpreted by reference to what I think, my interpreting my own thoughts makes no sense whatsoever. On the other hand, this means that the psychic elements constituting my actual thought do not stand in the same relationship to reality as do words."

Schrödinger then asked: "So, if thoughts are to give meaning to sentences they must have symbolic content. But this leads to a very vicious, almost insidious, type of infinite regress." Looking up at the wall over the fireplace, he continued: "For example, if I take down the plaque on the far wall commemorating Darwin, hold it up and utter a sentence, the sentence plus the plaque is capable of fewer and different interpretations than the plaque itself."

Absent-mindedly drumming his fingers on the table, Haldane noted that "What this all seems to add up to is that there *are* links between thoughts and language. But they do not require any inner vocalization. The question 'What are you thinking?' is not a description of an inner process, but an expression of my train of thoughts in words. Is that it, Wittgenstein?"

Before Wittgenstein could reply, Turing hurriedly continued this argument. "I see that the very capacity for having thoughts or beliefs requires the capacity to manipulate symbols. But this is not because unexpressed thoughts must be in a language; rather, it's because the *expression* of thoughts must be given linguistically."

Addressing Turing's point, Wittgenstein returned to his argument about making judgments. "Ascribing thought only makes sense where we have criteria for identifying thoughts. We must be able to distinguish thinking *A* instead of *B*. This means that thoughts must be capable of being expressed. And only a very limited range of thoughts can be expressed by the type of non-linguistic behavior one sees in, say, monkeys and birds. One should link thoughts more to potential behavior than to actual mental goings-on. But it certainly is the case that some sort of language is necessary. Human beings are essentially language-using animals."

"Well put, Wittgenstein. Aristotle himself could not have done better," observed Snow. "Let me see if I can summarize where we stand in this matter of language of thought. My sense of the discussion so far is that thinking is tantamount to making judgments. But to make a judgment requires use of a language. Consequently, *some* type of language is required for thought. And while thinking does involve a symbolic representation of real-world objects and notions, it is nothing so simplistic as a direct manipulation of those symbols in the brain in accordance with a set of linguistic rules; there is no *grammar* of thought. Finally, we have Wittgenstein's claim that human language is a rule-based activity, rather than a picturing relationship between objects and words. Now, have I left anything out?"

"That seems to cover just about everything," nodded Turing. "But in my mind it still leaves open the question of whether this language capacity can be given to a computing machine."

"Perhaps," replied Snow, "you are using the wrong verb. Maybe it's not something that's given *directly* to the machine, but rather something that the machine can *acquire*. After all, language does not get pumped into human children along with their mother's milk. It's something that's acquired by exposure to a particular community of humans using a particular language."

Turing agreed, noting that "It seems to me that an understanding of exactly how children acquire language would shed considerable light on the relationship between language and thought, perhaps even help settle the matter that Wittgenstein has raised about the distinction between language as a *tool* and language as a *picture*."

"Indeed," remarked Haldane. "It's patently obvious that if a computing machine is to duplicate human thought, then it's going to have to duplicate human language capacity as well. So understanding how humans acquire language should certainly give us some hint as to whether a machine might be able to acquire language in the same way."

As Simmons moved silently about from place to place, removing the salad plates and forks, brushing crumbs from the table and setting out plates for the dessert course, Snow thought for a moment as to which of his guests seemed best situated to summarize current thinking on the matter of language acquisition. Finally, he turned to Haldane and asked, "Perhaps as a biologist, Haldane is closest among us to the various competing theories on language acquisition. Would you be so kind as to outline for us the views of the scientific community at present in this area?"

"I don't mind giving it a go," answered Haldane, "but I want to make it clear at the outset that what I have to say is little more than an interested layman's account. I'm certainly not a linguist, nor have I studied the language acquisition problem in any great detail."

Having put forth these disclaimers, Haldane began his account.

"First of all, let's be clear on the empirical facts that any theory of language acquisition should be able to explain. These include the fact that every normal child is able to master a rich system of knowledge associated with *any* human language without significant instruction, that this mastery occurs even though the child is exposed directly to only a small fraction of all possible utterances in the language, and that language is acquired most rapidly when the child is between the ages of two and three."

Schrödinger interrupted Haldane's disquisition, remarking, "It seems clear that the most important of these facts is that a child can create statements that he's never heard before. This appears to be an extremely significant obstacle for any theory that depends on simple memorization or rote learning."

"Indeed it is," continued Haldane. "Strangely, however, current views on language acquisition suggest just the opposite; the 'poverty-of-the-stimulus' problem does not play a very crucial role in these theories at all."

"Could you please briefly explain one or two of these theories, Haldane?" entreated Snow, somewhat impatient to have him get to the core issues surrounding the language acquisition problem.

"I am sure that you have all heard of the school of psychological thinking called the *behaviorists,* or sometimes just *behaviorism,*" stated Haldane.

Schrödinger quickly remarked, "We noted earlier that they hold that it is unscientific to try to explain human behavior by postulating the existence of mental states of the brain."

"Just so. Some people have extrapolated this belief to mean that behaviorists claim that such mental states do not exist. But I sense that only a few very radical behaviorists would go this far; most simply say that if you want to create a scientific theory of behavior, then it must be based on directly observable phenomena, not on things like mental events that are, in principle, unobservable."

Haldane began outlining the behaviorist basis of human behavior, a theory that again called to mind the test Turing presented earlier in the evening for determining whether or not a machine is thinking. That test was completely in the spirit of the behaviorist program, since it was based on the notion that what is going on inside the machine is irrelevant insofar as deciding the machine's cognitive status; it is only the externally observed behavior of the machine that is involved in making this judgment.

"So what would a behaviorist say about the question of language acquisition?" Snow asked.

"Basically, he would say that learning a language is a kind of conditioned response, in much the same way that Pavlov's dog learned to drool in anticipation of food whenever a bell was sounded by its trainer," responded Haldane.

"So by this view a child learns the word *water*, for instance, by being given something to drink whenever it is thirsty. When this situation repeats itself many times, the child's brain builds up an association between the word *water* and this clear, cool liquid that quenches its thirst. Is that it?" asked Schrödinger, arching his eyebrows in faint disbelief, as if to suggest that only a psychologist would hold to a theory so at variance with the observed facts of language acquisition and performance.

"Essentially, yes," replied Haldane, in a somewhat sheepish tone. "I know it sounds incredible. But that's the heart of the behaviorist's view on language acquisition. It occurs by a long sequence of stimulus-response situations that become coded into the child's brain."

"Outrageous," exploded Wittgenstein, shaken out of his torpor by such an idea. "This view of language acquisition would imply that the child could never create new words or sentences that it had not heard from others. How could anyone believe that over the space of just a year or two a child could gain any degree of fluency in a language by means of such a learning procedure? Simply inconceivable!"

"I'm not defending the behaviorist view, Wittgenstein, just reporting it," reacted Haldane, rather defensively.

"Surely there must be other theories that do a better job of addressing the actual observed facts of language acquisition that you outlined earlier," said Snow, attempting to direct Haldane's attention away from Wittgenstein's angry criticism of the behaviorists and back to the topic at hand. "Perhaps a view of the mind as a rather less passive mechanism than these behaviorists seem to believe would lead to a theory more in accord with the actual facts."

"Indeed, there is a theory along these lines that has been proposed by a Swiss psychologist named Piaget," said Haldane. "He claims that the human mind generates what we call intelligent behavior as a process of reality construction rather than simply acting as a passive receiver and processor of information from the outside world."

Turing then noted, "As I recall, Piaget's theory differs markedly from the behaviorists in that he requires that an active, exploring mind makes use of internal mental representations in the course of its construction of reality."

"Right you are," continued Haldane. "Piaget thinks that introducing such entities into the study of the mind is no more unscientific than, say, a physicist like Schrödinger introducing an essentially unobservable particle like a neutrino into a theory of matter."

"And how does Piaget see the acquisition of language?" enquired Schrödinger.

"As I read him, Piaget says that language acquisition is no different in kind than the acquisition of any other skill at an appropriate stage of intellectual development. He feels that the human child goes through several phases of mental evolution from birth to puberty, ranging from the construction of concepts like space and time to the conception of the real world as a subset of possible worlds. Language is just one of the many mental concepts acquired during this overall process," explained Haldane.

"Just like learning to tie one's shoelaces or ride a bicycle?" asked Turing.

"No essential difference as far as I can see," replied Haldane.

"I'm afraid I still fail to understand how this theory of language acquisition can account for the way the child is able to be *creative* in the use of language," objected Schrödinger. "But perhaps by allowing—or even requiring—the use of mental concepts, Piaget can argue that creativity enters by means of assembling these concepts into various new types of patterns by some sort of internal grammatical shuffling around of linguistic symbols."

"Yes, neither of these theories does a very good job of coming to terms with the actual observed *facts* of how humans acquire and use language," admitted Haldane. "But they appear to be all we have, at the moment anyway."

During this interchange, Turing had been mumbling to himself and absently filling his note pad with arcane scribbles and doodles. At Haldane's admission that the theoretical explanations for how language is acquired were so feeble and did such a poor job of accounting for everyday observations that they didn't really add up to much, he interrupted Haldane's apologia to offer the group his own musings on the matter.

"We have agreed that language is one of the most characteristic features separating humans from other animals. And since language stems from the actions of our brains, it seems to follow logically that there must be something about the human brain that is fundamentally different from the brains of other organisms."

"That's pretty clear to all of us, Turing. But where does it lead?" asked Schrödinger, rather impatiently.

"Bear with me a moment," said Turing, as he continued his argument. "Suppose," he said, "there is a structure in our brains given to us by evolution that is specialized just for language, a kind of language 'organ', if you like."

Now came the punch line of Turing's theory: the notion that all human languages are at root the *same* language, in direct contrast to the prevailing views of linguists like Ferdinand Saussure and Leonard Bloomfield, who emphasized the diversity of human languages rather than their similarities.

"Let's make an argument analogous to the one I gave earlier about computing machines. In that argument I told you that the machine I sketched out is *universal*, in the sense that just by putting in a suitable program the machine could be made to emulate any other kind of computational process. Why couldn't it be the same with language? Maybe instead of human languages each being fundamentally different, as they appear on the surface, all languages are *exactly* the same in regard to their deeper structure."

Snow then asked Turing for clarification of this pivotal point in his theory. "Do you mean that just as all computing machines are essentially the same machine, which can be programmed to make it *look* like different machines, there is a language structure—I believe you called it an 'organ'—in the human brain that represents the deep structure of all human languages?"

"Indeed," replied Turing, warming to his theme. "Perhaps there is a part of our brain that is wired up specifically for language. Just as we can wire a computing machine to carry out the operations called for by any program, this language-specific part of the brain has the capability of implementing the specifics of *any* human language. All that's needed is for the environment to provide the details of the language that's to be learned. This basic structure then gets programmed by the environment so that the brain's owner here in Cambridge ends up speaking English instead of, say, the Chinese or Spanish that that person would have spoken had his brain been exposed to a Chinese- or Spanish-speaking environment in Hong Kong or Madrid."

"Are you claiming that acquiring a specific language like the German I acquired in Vienna is like giving a program to your computing machine?" asked Wittgenstein. "You seem to be saying that just as the program tells the universal machine what other machine it should pretend to be, the German-speaking environment of Vienna told the language organ in my brain what language it should implement."

Turing answered by elaborating on this idea: "What I have in mind is actually somewhat stronger than that. A couple of hours ago I explained how the material structure of a computing machine has to be able to carry out certain basic operations, things like moving a symbol from one location on the memory tape to another, or replacing a symbol with a different one. I think that this language part of the human brain is also wired in a special way, so that it can carry out certain basic operations associated with any human language. These might be things such as distinguishing verbs from nouns, stringing basic sounds together to create words, and putting words in sequential order to form sentences."

"Could you please tell us how this view of language acquisition helps explain some of the empirical facts Haldane mentioned earlier as to how children really acquire and use language?" asked Snow.

"Certainly the most important problem accounted for by this theory is what Haldane earlier termed the 'poverty-of-the-stimulus' problem. If every normal child has the basic grammar of all languages wired in to its brain from birth, then it's rather easy to see how even small children could form sentences that they have never heard before. Essentially, children already 'know' the structure of all possible grammatically correct sentences. You might imagine that every child has a template of every possible sentence structure coded into his or her brain, and all that needs to be done to form and utter an actual sentence is for the child to place appropriate words into the open slots in this template."

"Presumably, then, the words come from exposure to the child's native language, as does the activation of certain sentence templates and the rejection of others. Is that right?" asked Haldane.

"That's the general idea," confirmed Turing.

"This would appear to suggest the existence of a sort of *universal grammar* underlying all human languages," observed Schrödinger.

Snow enthusiastically joined in: "If Turing's theory is correct, there must be a 'deep structure' common to all languages, what Schrödinger has just termed a 'universal' grammar. However, what we actually observe in different human languages would then be simply a *surface* structure determined by the particular language being spoken. You've probably thought more about language than any of us, Wittgenstein. What do you say about this idea Turing has proposed?"

Shaking his head vigorously, Wittgenstein gazed up at the ceiling for a moment before shifting his attention back to the group. He began his attack on Turing's speculations by reminding them of his view of language as set out in the *Tractatus*. "Despite its attractiveness in addressing some of the most basic empirical questions about language acquisition, I think Turing's theory suffers from the same defects that caused me to reject my so-called 'picture theory' of language."

"How so?" asked Haldane.

"The basic problem with both theories is that they assume the human mind contains a kind of storehouse of symbols, with each symbol representing some kind of linguistic 'atom'. The theories then go on to suggest a type of logical calculus that combines these symbols in various ways to create a linguistic expression, which is then vocalized in spoken language. In Turing's case, this calculus is simply the inherent structure of this universal grammar; in the case of my own earlier theory, the calculus is a neurological implementation of the everyday logic of propositions. But as I've repeatedly said, there are fundamental flaws with this idea of a calculus. Language is much, much more than this."

Wittgenstein went on to re-state his position that language is a social phenomenon, berating Turing for ignoring this essential component of what makes language unique to human beings. Falling silent after this spirited and rather animated defense of his views, Wittgenstein withdrew into himself again as Schrödinger asked about the nature of this universal grammar.

"From what Turing has told us, it would appear that pure syntax is the essence of language, at least insofar as the inherent structure of language is encoded into this universal grammar in the brain. If this is so, then I see no distinction that matters between the kind of computer-programming language you outlined earlier, involving statements such as 'Replace 1 with 0' and 'STOP', and human languages. They would both be formal systems, essentially just a finite set of rules for manipulating symbols."

"Indeed they would," replied Turing. "The interesting question would then be what *kind* of formal system they are. Are they 'simple' systems having only a finite number of possible statements? Or are they infinite in their capacity? I would strongly suspect the latter. But in that case, there are many different sorts of infinities possible. And to understand language, we would have to know which of these possibilities is coded into the human language organ."

"But where is the *meaning* of a statement in your scheme of things, Turing?" Haldane said. "It's all well and good to talk about forming grammatically correct sentences using this universal grammar in the brain. But it's easy to make up such sentences that are utter and complete nonsense."

"For example?" said Turing.

"How about 'Red thoughts walk peacefully'," shot back Haldane. "There's an example of a proposition that is certainly correctly stated by the rules of English grammar. If your theory is right, then I was able to construct this sentence using the universal grammar in my brain that's been set to 'English mode'. Yet its semantic content is nil. So where does the fact that this utterance is entirely devoid of meaning enter into your theory?"

"I just don't know," admitted Turing. "What you call my 'theory' of language acquisition and performance is really just speculation at this point. It simply seems to me that the idea of such a language organ containing a universal grammar that's part of the birthright of every normal human child is a way of bypassing some of the obvious flaws in the theories of language acquisition you outlined for us earlier. But I certainly can't claim that my speculations are not without flaws of their own."

As Turing fell silent to brood on these flaws, Schrödinger added, "Why do we have to have it one way or the other? Why can't language acquisition involve both Turing's formal mechanism for syntax and a general learning capability for attaching meaning to the symbols? I don't see any logical reason why things couldn't be done in this fashion."

Snow saw that time was passing and this line of argument seemed to be veering off the main theme of what he'd brought his guests together to discuss. So he re-entered the conversation at this point and turned the discussion back to the issue at hand: the necessity for a true thinking machine to acquire fluency in a human language.

"We seem to be in agreement on the need for a computing machine to have human language if we are to regard it as really thinking." Glancing over at Wittgenstein, who continued to stare moodily off into the distance, Snow continued: "But there seems to be a division among us on exactly how a machine might acquire this linguistic capability. Turing argues for a language organ in the brain, that provides a universal syntactic structure upon which any particular language can be built simply by exposure; Wittgenstein, though, tells us that meaning is the essence of language, and that this can only be acquired in a social context."

Interrupting Snow's summary, Schrödinger noted, "Despite the fact that Turing's theory is based on syntax while Wittgenstein's rests on semantics, the two agree that the essential aspect of language acquisition is social; children learn a particular language by being exposed to a population of speakers of that language. So perhaps that's the point we should emphasize, rather than getting caught up in the details of exactly how the language is represented inside the brain."

"Just the matter I was coming to," said Snow, not without a touch of asperity. "Let's consider how a computing machine might be able to actually acquire the capacity for human language. Any thoughts on how this might be done, Turing?"

"I said earlier that one of the most encouraging reasons for believing that building a thinking machine is a viable proposition is that it is possible to make machinery to imitate any small part of a man. So if we are trying to produce a machine having human language, it stands to reason that one way to proceed is to follow the human model as closely as we can."

"Not *too* closely, I hope," commented Schrödinger wryly. "I'm not at all sure how eager I would be to have a computing machine curling up beside me on a cold winter's night."

"Hah!" snorted Haldane. "It'll be a *very* cold night indeed before anything like that's likely to happen to you, Schrödinger."

Embarrassed to the point of being red-faced by these allusions to Schrödinger's well-chronicled sexual proclivities, Turing quickly began explaining that that was not the kind of human mimicking he had in mind. But before getting very far with this clarification, he broke out into one of his stammering fits and was forced to pause for a sip of water to compose himself. After several moments, he got himself under control and continued.

"We should begin with a machine having very little capacity to carry out elaborate operations or to react in a disciplined manner to outside interference. Then by applying suitable interference—mimicking education, in effect—we should hope to modify the machine's structure until it could be relied on to produce definite reactions to certain linguistic circumstances. This would be the beginning of the process. How it might continue is very difficult to envision, at the moment."

"Have you carried out any experiments of this sort with the machines in your laboratory in Manchester?" asked Haldane.

"Actually, we have," Turing replied with a touch of pride. "The training of a human child depends largely on a system of rewards and punishments, which suggests that it should be possible to carry through the organizing of an initially unorganized machine with only two interfering inputs: one for 'pleasure', the other for 'pain'. We have experimented with a number of such pleasure-pain systems."

"Without going into technical details, I wonder if you could explain how these systems work?" asked Schrödinger, his eyebrows raised in curiosity.

"In general terms the systems are all structured in more or less the same fashion. The configurations of the machine are described by two expressions, which we call the 'character-expression' and the 'situation-expression'. The character and the situation at any particular time, together with the input signals from the environment, determine the character and situation at the next instant. Pleasure interference tends to fix the character, whereas pain stimuli tend to disrupt the character, causing features that had become fixed to change."

Snow noted that "This definition sounds too vague and general to be very helpful."

"Probably so," admitted Turing. "But the idea is that when the character changes we think of it as a change in the machine, but the situation is merely the configuration of the machine described by the character. The intention is that pain stimuli occur when the machine's behavior is wrong, while pleasure stimuli take place when it is right. With appropriate stimuli, one may hope that the character will converge towards the one desired and that wrong behavior will become rare."

Turing went on to describe the organization of one such system, in which the procedure was first to let the machine run for a long time with continuous application of pain stimuli, together with various changes of sensory data. Although the behavior of the machine eventually converged to one in which 'wrong' actions seldom, if ever, occurred, Turing told the group that the actual learning technique was really not very analogous to the kind of process that a human child would actually be taught. So he didn't believe that this particular experiment was quite the way a machine should be taught to acquire language capabilities.

"Let me see if I understand this experiment correctly," said Wittgenstein, in a deceptively gentle, almost subdued, tone. "You would give a machine all the sensory apparatus of a human being—eyes, ears, a nose for smelling and skin for touching—and place this robot into a human environment. By being exposed to the language of this human community, you then believe that this machine would acquire language by reconfiguring the electronic circuits making up the robot's brain. Is that your game, Turing?"

"More or less," responded Turing.

"This is completely preposterous," exclaimed Wittgenstein. "One can ascribe thoughts—including meaningful linguistic expressions—only to creatures who are engaged in a way of life in which those expressions make sense. The concept of 'pain' is characterized by its particular function in our life. We only call a feeling 'pain' that has this position, these connections to our life. You might label your machine's behavior as *pain,* but it has no more connection to what we humans call 'pain' than if you'd labeled it 'pleasure', 'humor', 'joy' or 'grief'. The machine cannot *understand* what you mean by 'pain' any more than it can understand these other human emotions. And without such understanding, a machine certainly cannot be said to know language."

As Wittgenstein ended this outburst against the possibility of a machine ever acquiring true human language capabilities, Simmons appeared in the doorway and discreetly signaled Snow that he was ready to serve the dessert course. Seeing that the competing views on language acquisition by a machine had begun to devolve from actual facts to vague, personal opinions, Snow thought this was an opportune moment to redirect the discussion into more productive channels.

"Gentlemen, I see Simmons beckoning from the doorway. Perhaps this is a signal to turn our attention away from language to another, equally fascinating, human cultural construction: the culinary arts. So may I suggest we take a pause and let him clear away these plates and serve the dessert? Both Turing and Wittgenstein have suggested that the isolated man does not develop any intellectual powers, linguistic or otherwise. So when we reconvene, I would like to take up the development of other aspects of human culture besides language, things such as religion, art, literature and other creative artifacts. I have a suspicion that such a discussion will shed a different type of light on the issue of whether machines could ever acquire the type of intelligence we associate with humans."

Chapter Six

The Dessert

Life and Personhood

Wheeling in a dessert trolley, Simmons gave everyone a tall glass filled with boiled oatmeal and fruit, garnished with a generous dollop of whipped cream, the whole thing topped off with a tot of single-malt Scottish whisky.

"Aha," exclaimed Haldane, clapping his hands in appreciation. "A flummery, Scottish style. I simply must compliment you, Snow, on choosing the perfect dessert to top off this very excellent meal."

"Perhaps it will sweeten up our discussion a bit," Snow replied drily. "Our deliberations have centered on the role of language in cognition, but I wonder how social values and culture fit into all this. If language is so important to human thought, then broader cultural matters must be also. After all, language is just an expression of a culture, albeit a singularly important and visible one."

"Indeed, it is" added Schrödinger. "I believe it was your British writer George Orwell who once said that

'Political structure determines language and language determines thought.' So if he's right, and we are right in focusing on language as the central ingredient in human thought, then any discussion of machines thinking like humans is going to have to take cultural and social factors into account."

Pushing his flummery aside, Wittgenstein replied to these arguments, saying, "Take this strange dessert that's just been put in front of us. Haldane says it is a Scottish concoction. Now what makes him say that? Why is it immediately recognizable to him as Scottish and unrecognizable to Schrödinger or myself as anything other than a mixture of grains, fruits, cream and whisky? It's simply because Haldane is familiar with Scottish culture, and so he sees his glassful of this mixture as being Scottish; I, on the other hand, have no such picture in my mind. So for me the thought of Scotland and Scottish whisky distilleries never enters my mind. What we see depends on where we've been and on the totality of our life experiences. So for one of Turing's machines to fool me into thinking that it's Haldane, it would have to have had the *same* life experiences as Haldane. For thought, language matters and language is just the expression of a culture in words."

Before Wittgenstein could launch himself fully into this theme, Schrödinger interrupted, saying, "This argument seems to suggest that a population of intelligent machines of the type Turing proposes would have to develop all the cultural traits we see in human populations. Or at least they would have to develop things like religion, art, language and so on if there's to be any chance that we would consider their intelligence to be 'human'."

Rising to the bait hidden in this remark, Turing re-emphasized his contention that these intelligent machines must be given the very best sensory apparatus that money can buy if they are to explore the physical world and learn about it, just as humans do.

"But to speak of a *single* machine, or 'robot', of this type is a quite different matter than to talk of a *population* of such devices," said Snow. "Initially, I thought we were talking about the first case. But since we've been arguing language capacity and culture as being hallmarks of human intelligence, we seem to have moved into the latter. For myself, I can't see how one can meaningfully speak about culture arising on a desert island populated by a single human being—or a single robot either, for that matter. What do you think, Haldane? You're the expert here on how organisms arise and populations evolve."

"Some years ago, I suggested that the origin of life is to be found in large, organic molecules that were likely to be plentiful in the oceans of the early Earth. Moving about in this dilute organic 'soup', these molecules could interact by a process analogous to crystallization, and reproduce their kind by building up similar molecules from simpler constituents. Life, then, would only begin when a number of such molecules were more or less permanently associated with each other."

"By this notion," said Schrödinger, "if you were to break up a bacterium and pass its component pieces through a filter, it would cease to be alive. But when the filterable materials are again built up into bacteria, life would then perhaps begin anew. Is that right, Haldane?"

"Precisely," said the biologist. "Life is a resonance phenomenon between molecules."

Turing then added: "Most higher plants and animals can be subdivided to some extent without killing them. But a living cell cannot be divided like this;

cutting it apart destroys its functioning as a living object. This would appear to suggest that the cell is a far more fundamental unit of life than the whole, multicellular organism."

Warming to this theme, Haldane responded with gusto: "Exactly so. This is just what one would expect if life is a phenomenon of resonance between molecules rather than larger structures. As I recall, Schrödinger, you have expressed much the same views in your little volume *What is Life?*, which seems to be receiving considerable attention nowadays."

Before Schrödinger could reply, Wittgenstein broke in, saying, "Let's just suppose you are right and that living processes did get their start in your primeval 'soup'. What about the mind? When did a simple bacterium evolve into something that we could say had a mind?"

Haldane replied by arguing that mind is also a resonance phenomenon, just like life. But a resonance between collections of objects larger than simple atoms, perhaps aggregates of cells undergoing periodic electrical disturbances.

"But there is no reason to attribute mind to such organisms as protozoa or higher plants, and it is probably at best exceedingly rudimentary in all but a few animals," Haldane went on to state with some deliberation. "My belief," he said, "is that mind is not some kind of mysterious phenomenon that is piled on top of matter, but is a separate entity interacting with ordinary material systems."

Rising up from his chair to dispute this point, Turing noted that such a view of mind would be tantamount to claiming that the study of mental phenomena was no longer the province of science, since Haldane's position relegated the existence of mind to a question of metaphysics.

Sensing the discussion wandering off into a lengthy digression on metaphysical speculations about minds, Snow cut off Turing's objections before they could be fully developed. "Let's think about how all this talk of the origin of life and mind fits together with the question we are here to discuss," he said. "What does it have to do with whether we can create a machine that thinks?"

"To see how life and mind—or, if you like, living and thinking—are intertwined, we first have to get a better grasp of what it is that separates living from non-living things," responded Schrödinger. "In what way *exactly* do we regard this fly buzzing around the table as being alive and not grant that very same property to this chair I'm sitting on?"

"I'm sorry, Schrödinger, but it's just not clear to me why we have to define life before we can proceed to consider the possibility of a machine that thinks. What does the definition of life have to do with this question?" asked Snow.

"We have all agreed, I believe, that in order for a machine to display the full spectrum of human intelligence it would have to possess the full spectrum of human sensory apparatus so that it could interact with the world and learn about human life, in much the same way that human children do," said Schrödinger by way of reply.

"That does seem to be our common position," Turing agreed.

"Well, then," continued Schrödinger, "how can you have all these sensory and information-processing capabilities leading to cognition, and ignore the issue of precisely what kind of objects can possess them? As far as we know, only living things combine sensory and information processing to a degree giving them cognitive awareness. So perhaps it would be a useful step to ask what is alive and what is not."

"I see now what you're driving at," Snow conceded. "Perhaps it would sharpen the issue if you would give us some indication of your own views on the 'fingerprints' characterizing a living organism."

Schrödinger was happy to oblige. "In my recent lectures in Dublin, I focused on the information-carrying capacity of living things. In particular, my concern was with where *exactly* the living cell stores the information needed to make a copy of itself to continue the life of the organism. My tentative conclusion is that this information is stored in a non-repeating crystalline pattern in the proteins making up every living cell."

"But why do you place such emphasis on this question of information storage?" asked Snow.

"Principally because the primary functional activities characterizing living organisms all depend on the availability of information. So the key element is where the information is stored and how the cell makes use of it."

"What types of activities do you believe distinguish living from non-living things?" asked Turing.

"To my eye, there are three features that separate you, Turing, from a stone on the street. The first is that you have a *metabolism,* by which you take energy from the environment and process it to enhance your own survival. Next, you have procedures built in to your cells for *self-repair* when the cellular machinery begins to run adrift. And, finally, your cells are able to *replicate,* in that they can manufacture good—but not necessarily perfect—copies of themselves."

"So are you saying that the functional activities separating living things from the non-living are metabolism, self-repair and replication?" asked Haldane.

"Precisely," confirmed Schrödinger.

At this juncture, Wittgenstein interrupted Schrödinger's lessons on life by asking: "You can't seriously be suggesting that simply by possessing these properties of life a machine would all of a sudden acquire human cognitive abilities?"

"I am suggesting no such thing," Schrödinger shot back. "These properties are necessary for life. But possessing them is certainly very far from saying that an organism is in any way cognitively similar to a human being."

"I should say so," Haldane snorted. "Even a lowly bacterium has the features Schrödinger claims characterize the living state. And I certainly wouldn't want to have my thought processes likened to those of a bacterium. Nor would any of us here, I'd wager."

Snow stepped in at this point to direct the discussion back to the main theme, of what kind of properties a machine would have to possess to be regarded as a thinking being. Biology is all well and good, he thought, but there's more to being a human than mere organized chemistry. What about morals, ethics and even the issue of identity? What does it mean to be a person? he asked himself. And what does personal identity have to do with thinking like a human? This seemed to him to be a point worth putting on the table for the group to consider.

"Let me shift the discussion from biology to what I suppose can only be considered as philosophy," he began. "I would like your collective wisdom on the matter of personhood for machines. Suppose the engineers and other boffins actually managed to construct a device with sensorimotor apparatus of the sort Turing wants, and even endow this machine with all three of Schrödinger's qualities of life. Could such a machine

141

then be expected to 'grow' into a cognitive being that we would eventually consider to be a 'person'? Or is this some kind of anthropomorphic fantasy?"

Almost before Snow could get these words out of his mouth, Wittgenstein shot forward in his chair like a man poked with a sharp stick, exclaiming, "It's absolutely unthinkable, Snow, to suppose that a machine— even one with all the fantastic properties you have given it—could be considered a 'person'. The whole idea is a gross confusion of categories. It would be analogous to speak of the number 7 as being 'green'. It's simply not possible to conceive of personhood for any object other than a flesh-and-blood human being."

In no way cowed by this typical Wittgensteinian outburst, Haldane retorted: "Hold on a moment, Wittgenstein. I think we have to consider the issue of what we *mean* by personhood for humans before we can jump to the conclusion that a machine could not have it too."

"I agree," chimed in Turing. "Without considering what it means to be a person, how can you say that a thinking, living machine of the sort Snow described cannot be one? It's just plain wrong-headed reasoning to assert this by fiat, not to mention being bad philosophy as well."

"So," said Snow, "let us indeed consider what it means to be a 'person'. Do you have some thoughts on the matter, Schrödinger?"

Shifting around in his chair, Schrödinger idly toyed with his dessert spoon as he looked over at Snow in a faintly bemused fashion. After several moments of silence, he haltingly began confronting the challenge that Snow had set for him, to give an account of what it means to be a person.

"Let me say at the outset that I see this issue as divided into two parts. First, there is the problem of *personal identity.* Just what is it that allows us to say a person is the *same* person over time? This leads directly to the notion that we've been considering up to now, namely the conception of personhood by persons."

"But how can personal identity be a problem?" wondered Turing. "Isn't the identity of a person over the course of time one of the most fundamental truths of all? Just think of all the implications a change in this view would have throughout our entire system of beliefs."

"Indeed," Schrödinger replied. "This is precisely why we have to have the notion clear in our minds before we enter into a discussion of this kind. Otherwise we are in a muddle that leads nowhere. So just what is it *exactly* that enables me to say that the Wittgenstein I see here and now is the same person that I came up the stairway with a few hours ago?"

"Or, for that matter, the same Wittgenstein you met in Austria a decade ago," added Haldane.

"If memory serves me correctly," broke in Snow, "there are two traditional theories on this. The first says that a person is above all a physical organism in a fluctuating state. So the identity of the person amounts to the identity of that physical system. We might think of this as the 'body' half of Descartes' mind-body dualism. The other theory claims that the continuity of mental states is what counts insofar as being a person. This is the 'mind' half of personhood."

"So the physical theorist argues that if Snow here were a short, thin Chinese man when you last saw him, then if a tall Irish woman came to the door claiming to be Snow we would have no reason to take her claims seriously. Rather, you will say that it is impossible and slam the door in her face. Physical organisms are just what persons *are.* Is that it, Snow?" asked Haldane.

"In your inimitable style you have indeed captured the essence of the physical theorist's position," replied Snow.

As Wittgenstein cringed slightly at this interchange, Turing enquired about the mental theorist's view. "My understanding is that the mental theorist says this kind of physical argument confuses an operational principle with a theoretical insight. Instead of focusing on physical structure, the mental theorist would say that it's perfectly possible to imagine awakening one morning and discovering that one has an entirely new body."

"In fact," noted Schrödinger, "this was just the premise of Kafka's famous story, 'The Metamorphosis', in which the hero, Gregor Samsa, awoke one morning to discover that he had been transformed into a giant insect."

"An instructive example, to be sure," continued Turing. "Now it seems to me that if we can believe in the very possibility of some such thing happening, then our notion of what constitutes a 'person' is not completely linked to our concept of a person's physical body. If it were, this Kafkaesque scenario would be quite unthinkable."

"But what do you understand by 'mental continuity'?" enquired Schrödinger.

Recalling that the philosopher John Locke had suggested that memory is the relevant kind of mental continuity, Snow argued that it is memory that links the past experiences of an individual to his present consciousness and in just such a way that they cannot be linked to any other individual's consciousness.

Able to remain silent no longer at what he saw as a completely misguided discussion, Wittgenstein leaned across the table, fixed Snow with a steely-eyed glare and asked: "What then if you had a 'mind-erasing' technology, so that you could wipe an individual's memory clean without affecting his physical health. Suppose that society uses this erasure of memories as an alternative to capital punishment. Now I ask you, Snow, which would you prefer: to have an organic death by the customary barbarities, such as the hangman, or having your mind erased?"

Peering over at Wittgenstein from beneath his heavy horn-rimmed glasses, Snow replied, "I don't see much of a reason to choose one over the other. How would I be better off by having my brain erased?"

"That sounds like a backdoor argument in favor of mental continuity theory," noted Turing. "What you seem to be saying is that physical continuity has little value, if mental continuity is destroyed."

"Just a minute," interrupted Haldane. "Before we jump to this conclusion let's look a bit harder at the physical continuity theorist's position."

Haldane then noted that it is not at all uncommon for parts of a person's body to be missing without in the slightest affecting our perception of an individual's personhood. "People lose a leg in a war or have an infected eye removed without our thinking in the least that that individual has become a different person. It only seems to be the brain that counts as far as personal identity is concerned. So, it is brain continuity, not overall physical continuity, that supports the mental continuity theory. In this sense mental continuity is more important than physical continuity after all."

"Well," said Turing, "if only the brain is to count, let me propose the following thought experiment. Let's take a detailed map of all the neurons, synapses and every other connection in Haldane's brain. Now suppose I replace each of these components with an electronic counterpart—vacuum tubes, wires, resistors and so forth—all connected up in exactly the same way as the components are linked in Haldane's brain. Then not only should this electronic copy of Haldane's brain think *just like* Haldane, by the physical continuity theory it should *be* Haldane."

Schrödinger then added: "By this amazing fantasy, Turing, you are in effect asserting that Haldane's mental states are the result of computational powers of the brain. This is a kind of *computationalism,* in which those mental states are the process of the brain, which leads to personal identity as being simply 'process identity'."

Responding to Schrödinger's shrewd observation, Turing stated, "If what you call computationalism is true, it is in principle possible for me to transfer the functional state of my brain at some moment to a computing machine, and ask that machine to carry on my brain's processes."

"So," said Snow, "when this transfer is complete and the processes of your brain are 'reactivated', this new system would be Turing in every possible way other than the purely material. The machine's mental states would simply pick up where Turing's left off, in much the same way that my mental states pick up in the morning where they left off when I went to sleep the night before."

Wittgenstein again rose to the challenge, objecting that, "This computationalism offers nothing less than a kind of immortality. Once 'you' are captured in a program and a set of data records, you will not perish as long as there continue to exist computing machines complex enough to receive you and execute your program. What astounding nonsense this all is!"

"What do you mean?" Snow asked.

"I mean that suppose Turing here decided to copy his brain into a machine at the moment of his death. But instead of copying it into one machine, some clever wag decides to play it safe and arranges a simultaneous copy to be made into *two* machines. Now all of sudden we have two Turings that are functionally and mechanically indistinguishable."

"Well, so what?" said Haldane. "So much the better, I'd say, since now Turing will have two bodies instead of one and could then experience twice as much."

Throwing up his hands in frustration, Wittgenstein flopped back into his chair with a look of disgust on his face. But before he could reply to Haldane, Schrödinger interrupted, saying, "No, that's no good. I see Wittgenstein's point—and it's a subtle one. Turing had planned on only one copy, and that copy would *be* him. But now there are two copies who cannot both be Turing. To think otherwise is a gross violation of basic logic. For suppose that T is Turing before his death, while A and B are the two copies. Then by Wittgenstein's scenario, A and B would have to be one and the same individual. But that makes no sense, since A and B are two distinct individuals, regardless of how much they resemble each other."

"Yes, indeed. I see the point now," acknowledged Haldane. "A computationalist would claim that A is T. And whatever justification there is for this claim is equal justification for saying that B is T. But there can be no argument for asserting that A is B. Well, maybe this just shows that computationalism is wrong."

"I'm not so sure," Snow said. "This kind of division problem is not confined to computationalism, but pervades any account of personhood that explains mental continuity on the basis of continuity of a process rather than an entity."

Snow went on to explain the difference between dividing something like a cake and an event like a concert. In the first case, we can share the cake by dividing it into slices, and in so doing reduce the physical nature of the cake. But suppose we have a musical performance taking place in a concert hall, and that the performance is broadcast over the radio so that many people can share it by tuning in to the broadcast. The event of the concert is then 'divided'—but not in the sense that two people divide a cake. These two notions of division, Snow said, only become puzzling when we speak of personal identity because we are accustomed to thinking of ourselves as entities, not processes. So this kind of argument really goes in support of the physical continuity theorists.

"After hearing all these arguments," said Schrödinger, "I'm coming to the view that perhaps what matters in all this deathbed transference is not survival of Turing, but rather simply continuity. We have been focused here on whether Turing, *as Turing,* survives this transfer to a machine. But this leads to all these logical problems about whether what survives is really Turing or not. Better in my view to say that what matters is whether there is *anybody* who is Turing's continuant."

"This sounds rather close to the Vedantic view of the Eastern mystics," noted Haldane.

"Precisely," Schrödinger replied enthusiastically. "Let me explain this by posing four questions that bear directly on what we have been discussing. Does an *I* exist? Does there exist a world besides me? Do *I* cease to exist on the death of my body? Does the world cease to exist on my bodily death? My answer to all four questions is to say that there exists only one universal being, which the Vedantic tradition calls Brahman. And Brahman encompasses all reality in a seamless, undivided unity; Brahman is pure thought."

"I recall discussing this tradition during my visits to India," noted Haldane. "By my recollection, Brahman is associated with a power called Maya to which is due the appearance of the world; Maya is the so-called material cause of what we sense as the world."

"That's correct," Schrödinger replied. "The unenlightened soul is not able to look beyond Maya. But the enlightened soul, which is really Brahman, is enmeshed in the unreal world of Maya, and the only way out is by the Veda. The self and world are one—and they are all. All selves are united in one consciousness; we are all different aspects of a single mental unity."

Snow then noted that Schrödinger's 'single unity' view was in direct contrast to Wittgenstein's earlier assertion that the fact that we can share a meaningful language and beliefs is a social fact; words are not the personal property of isolated individuals, but draw their primary meanings from the public uses that they are given in collective situations.

"Indeed," added Wittgenstein, "the public character of the way in which words gain their meaning throws considerable doubt on the image of individual human experience that Schrödinger seems to regard as being inescapable, that of hermetically separated consciousnesses."

Schrödinger responded: "Let me use one of your own arguments from your earlier work, Wittgenstein, in order to support my position. The solution of this plurality versus unity dilemma cannot be logically proved, since logical thought is part of the phenomena and wholly involved in them; it can only be *shown* pictorially. These common experiences that you speak of that underlie the meaning we give to words lead to relationships that have never been grasped by formal logic or exact science. These relationships lead us back to metaphysics, to something that transcends what is directly accessible to experience."

"Perhaps this metaphysical note is a good place to draw this topic to a close," said Snow. "I don't see that there is any rational way of settling this person-hood argument, at least not here tonight. So let me shift gears just a bit and ask you all to consider the notion of social behavioral patterns that seem so char-acteristic of humans. I'm thinking of both individual social actions such as the creation of artistic works and altruistic acts, as well as group activities like religion and political structures. These also seem to be part of what it means to be human. So I'd like to hear your thoughts on what they may have to do with duplicating human thought in a machine. For example, could we ever expect to see something like a 'machine culture' emerge from, say, copying human brains into individ-ual machines?"

"I say, Snow," interjected Haldane. "Before taking up such weighty matters, could we perhaps take a small break and then reconvene in the drawing room for this discussion over a spot of cognac and a good cigar?"

"A first-rate idea, my friend. I'm sorry I didn't think of that earlier. But yes, let's leave this table to the minis-trations of Simmons, and stretch our legs for a moment before moving into the drawing room to continue these speculations."

Chapter Seven

The Cigars and Brandy

Social Behavior, Culture and Thought

As Simmons made his way around the drawing room, offering coffee, Snow leaned back against the fireplace mantel, a snifter of brandy in his hands, and said to the group: "May I suggest you try this excellent brandy? And please help yourself to the cigars on the side table. They're Cuban Montecristos. Repayment for a small favor to a friend in the Foreign Office."

"So, despite their best efforts, the Foreign Office has some redeeming merit, after all," remarked Haldane, as he went through the customary rituals of moistening, clipping and lighting his cigar.

Wittgenstein and Turing took up places at opposite ends of the large sofa, each looking rather morose and introspective. Schrödinger spoke from the easy chair directly across from the fireplace. "You know, Snow, we have spent considerable time this evening discussing the social component that seems to underlie what we humans call 'intelligence'. I wonder, if one had a computing machine of the type Turing is proposing, complete with the best sensory apparatus that money can buy, whether a population of these machines would form themselves into social units of the sort human beings and other animals have built for themselves. Or is this notion too far-fetched to be taken seriously?"

Re-energized by Schrödinger's bold, visionary speculation, Turing immediately responded: "In America last year, Professor von Neumann showed that there is no logical barrier to the existence of a machine that could make copies of itself. So if one imagines a population of such self-reproducing machines, then it's only a small step from there to suppose that natural selection might lead to the formation of social groups, and even a kind of 'culture', emerging in such a population of machines."

"Tell us more about von Neumann's result," said Schrödinger.

Turing went on to describe what von Neumann had in mind. "He envisioned a machine set adrift amidst a 'sea' of raw materials from which the parts needed to reproduce itself can be fabricated. Von Neumann then showed that if this machine contained a blueprint specifying its own plan, had the capacity for universal construction and had a control unit and a copier, then it would be logically possible for such a machine to manufacture a perfect copy of itself."

"And how does the machine accomplish this miraculous task?" asked Haldane.

"The basic idea is that the machine first reads its blueprint and constructs a copy of itself by assembling the various parts needed from the environment. But this copy does not yet contain the blueprint; it is simply a copy of the machine. The control unit then shifts from 'construction mode' to 'copy mode', using the copier to simply duplicate the blueprint. This copy is then attached to the new machine, thereby producing an object that is now a *complete* copy of the original—including the blueprint."

At this point Haldane again interrupted Turing's account of von Neumann's work, saying "Ah, but the evolutionary forces leading to a new species require just the opposite. The copies must be *imperfect*, so that some copies are created just a bit more equal than others. It is this difference that gives natural selection something to grab on to and work with."

"Well, I suppose one could see how this might come about during the process of reproduction," replied Turing. "Suppose that when the machine fabricates a grasping tool for its copy's 'hand', it makes a small mistake by cutting the 'fingers' to the wrong length. Such a grasping tool would still function, and, in fact, it may even function better than the original. So in this case the copy would be just a bit more capable of surviving in its environment than the original. That way, natural selection might act to promote more copies of the copy in future generations than of the original."

Snow then remarked, "Of course, that could only happen if the mistake in the cutting could somehow be incorporated into the actual plan for the machine that is built in to the copy. Isn't that right?"

"Or errors could be made directly in the blueprint," answered Haldane. "One way or another these sorts of mutations have to find their way into the plan if they are to show up in future generations. This is the basic thesis of evolutionary genetics, and is what has caused all the trouble recently in Russia over the ideas of Lysenko. This charlatan has somehow convinced Stalin that traits like longer fingers that are advantageous in the environment can be passed on to the next generation *without* being coded into the organism's plan."

"Do you mean this Lysenko business going on now in Russia is all over simply the kind of inheritance advocated years ago by Lamarck?" enquired Snow.

"Indeed, that is the gist of it," replied Haldane. "But the political overtones have swamped any scientific aspects of this whole business."

Schrödinger then asked: "Perhaps you could say a bit more about Lysenko and why his ideas on genetics have caused such a furor in the scientific world?"

"Trofim D. Lysenko," explained Haldane, "is an uneducated peasant who has managed to almost single-handedly destroy Soviet agriculture—and much of Soviet science—for nearly a generation. He was a farmer's son who in 1929 tried to tell the world about his experiments in growing winter peas to precede a cotton crop. He thought this discovery was sensational—but it was not sensational at all. In fact, the idea was a very old one."

"What happened?" asked Turing.

"A year later Lysenko's father sowed grain in winter and got a yield in the spring. Hearing about this, Lysenko immediately claimed credit for the idea, saying it proved his own agricultural 'theories'. His loud bragging paid off by helping him land a position at the Odessa Institute of Genetics and Breeding. And since winter crops were notoriously poor, Lysenko was put in charge of a special department to study this problem.

"In 1935," Haldane continued, "Lysenko gave a talk at a meeting of representatives from collective farms, the essence of which was that those who didn't agree with his theories of plant breeding were enemies of the people. Quite by chance, Stalin himself was in the audience and afterwards commented, 'Bravo, Comrade Lysenko.' From that moment onwards, everyone assumed Lysenko was a protégé of Stalin's, and he was immediately lionized in the Soviet press as a 'genius of the soil' ."

"Ah, the political dimension enters the picture," nodded Schrödinger.

"Yes," replied Haldane, "and shortly thereafter, Lysenko acquired a coterie of followers who were given titles, awards and top jobs in all aspects of Soviet scientific life, as well as on the editorial boards of newspapers and magazines. Strange business. It's becoming apparent that biology and genetics are developing greatly elsewhere in the world. But Lysenko still has Stalin's ear, and is misinforming 'The Leader' as to the state of all science in the USSR. I can't begin to count how many honors and titles the man has been given. They've even erected statues of him in some provincial towns."

"But is there any scientific basis at all to Lysenko's theories?" asked Snow.

"You see, Snow," replied Haldane somewhat bitterly, "Lysenko's scientific ideas are nothing but a load of rubbish. One might as well argue that dogs living in the wild give birth to foxes. I can't think of a geneticist alive who would give even a moment's consideration to such nonsense. And the worst part of all this is that Stalin won't listen to anyone but Lysenko on these matters. They fit so well with Marxist ideology. Let me tell you, many fine Soviet geneticists have been arrested— and even shot—for speaking out against Lysenko."

"How did you get caught up in this mess?" asked Schrödinger.

"Somehow the idea got around that since I'm both a geneticist and a Marxist, I must support Lysenko's crackpot theories. In fact, it got so bad that I was forced to publish an explicit condemnation of this rumor earlier this year in *Modern Quarterly*. It's a sad day for the socialist movement when political ideology starts dictating to nature what she can and cannot do. Makes one rethink one's commitment to the cause, eh?"

At this point Wittgenstein felt compelled to caution that "Whatever you might say about the Lysenko business, it is certainly not Western. So you cannot measure Lysenko's reasoning by Western standards. But the worship of science is the greatest evil of this century. So for this reason alone I am opposed to Marxism."

Turing added, "Marxism makes claims of being scientific—but only as an expression of a need for a rationale of historical change that can be justified by science. But how can anyone seriously entertain the Marxist notion of explaining something like science by 'prevailing modes of production'. It's total nonsense."

"Not necessarily so," retorted Wittgenstein. "Who knows the laws according to which a society develops? I am quite sure they are a closed book to even the cleverest of men."

At this strange remark by Wittgenstein, Snow intervened: "Perhaps we can get back to the business at hand. What can be concluded from all this about the likelihood of a population of intelligent machines evolving social notions like a political system or, for that matter, even cultural constructs like religion or art?"

Rather annoyed with this question, which seemed to suggest that there might be no difference that matters between humans and machines, Wittgenstein leaned forward and answered sharply: "Look here, Snow. These notions of culture you're talking about are completely tied up with the idea of being human; they are part and parcel of the flow of human life. Now I suppose it's possible to conceive of these machines as sharing a form of life of their own. I don't know. But even if they do—and even if this form of life leads them to develop something like a 'religion'—it certainly will not be the same as what we humans mean by a form of life. A machine religion, for instance, would be at most a kind of pathetic parody of human religion."

"I'm not at all convinced of that," said Schrödinger. "In the Eastern religions there is often very little distinction drawn between animals and humans insofar as both are regarded as part of a single unity—the Veda, for instance. So it seems to me that to draw a distinction solely on the basis that humans are carbon-based entities while machines are made of different 'stuff' is questionable, to say the least."

At this introduction of non-human animals into the discussion, Snow enquired: "You may see a centipede as having a soul, Schrödinger, but this is not really an objective claim. It strikes me more as a matter of faith or even an appeal to a sectarian religion."

"I beg to disagree," said Schrödinger apologetically. "This is not at all a religious view—Eastern or otherwise. For instance, Leibniz argued in the seventeenth century that everything, man or rock, was composed of an infinite number of tiny souls. And I think Leibniz's contemporary Benedict de Spinoza would have agreed with him."

"But does having a soul imply a mind?" asked Haldane.

"Not at all," answered Schrödinger. "The point I'm making is that the idea that things like minds and souls are by custom reserved for human beings is sheer anthropomorphic chauvinism, and there is really no logical reason why they cannot be possessed by animals, stones, clouds or, for that matter, machines."

"Well, what about normal human social behavioral patterns, such as being nice to your relatives or avoiding things like incestuous relations? Can we really expect a society of machines to adopt such cultural norms just because they are intelligent and think like humans?" asked Haldane.

With this query, Haldane put on the table the question of the influence of culture on behavior as opposed to behavior dictated solely by inheritance.

"The issue of how much genetic inheritance determines human behavior has been debated almost from the time of the publication of Darwin's pathbreaking work," noted Snow. "So to some of us present, like myself, debating this question in Darwin's old rooms at Christ's seems strangely appropriate, although not without a certain touch of *déjà vu*."

Stepping cautiously into this intellectual minefield, Schrödinger offered the view that "If we accept our own animal nature, then it seems quite reasonable to suppose that many of our behavioral patterns are simple consequences of this fact. I would imagine that a society of machines would develop analogous patterns from their own evolutionary history. But I do not see any reason why those patterns should necessarily mimic the ones we have inherited from *our* evolutionary past."

"Let's try to separate this question into its component parts," said Snow. "On the one hand, there is the issue of what we might term 'nature versus nurture' in humans. Then there is the completely separate question of whether the evolution of social behavior in machines would follow that of humans. I see these as two very different questions that should not be conflated."

"Indeed they are," said Haldane. "When it comes to human behavioral patterns," he went on, "certainly one of the most puzzling types of behavior for genetic determinists to explain is the act of being kind to others at your own expense. If our behavior was really aimed at enhancing the ability of an individual to place as many of his or her genes as possible into the next generation, it's difficult to see how sacrificing yourself to aid someone else could contribute to this goal. So perhaps our cultural norms rewarding this kind of behavior outweigh Darwin's 'survival of the fittest'—at least when it comes to human behavior."

"I'm not so sure," objected Turing. "There are several ways one might explain altruistic behavior in terms of Darwinian fitness without having to invoke some special culturally learned actions peculiar to humans."

"For example?" asked Snow.

"Well, one way would be for closely related family members to help each other. Since two brothers, say, share half their genes, if one brother helps the other, this action has the effect of raising the likelihood of the genes of both brothers surviving into the next generation."

"I see what you mean," replied Snow. "In fact, the two parties wouldn't even have to be related, since if I help Haldane's genes survive such an action can also help my genes get into the next generation—provided that I can *expect* Haldane to repay my assistance in kind."

"Yes. This would be a kind of 'reciprocal' altruism," Turing noted.

"All right, we can see how seemingly altruistic behavior in humans might be explained by purely genetic self-interest. But what about machines?" asked Schrödinger. "After all, that's what we're really discussing here."

"Just so," agreed Snow. "What about machines?"

Turning away from the window, where he had been staring out at the moon that was finally beginning to peek out from behind the clouds, Wittgenstein strode to the center of the room and exclaimed: "You're all talking rubbish. You seem to think that there is no difference between some kind of intelligent machine and a man, other than the 'incidental' fact that one is made of flesh and blood while the other is composed of metal, glass, wood or God knows what else. Well, I tell you that there is an unbridgeable gap between the two—and it has nothing to do with the matter from which they are composed. Humans are humans and engage in human social interactions; this is what makes them human. Machines, no matter how cleverly they imitate humans, cannot *become* human simply by imitating human social interactions. Yes, they may have machine dreams—but those dreams are as far from being the dreams of a human as a steam shovel is from being the college gardener digging in the courtyard. It's a fundamental category mistake to even talk of machine thoughts, societies and the like in the same terms as we discuss those same concepts for humans."

Collapsing back on to the sofa after this impassioned statement of his skepticism at the very idea of creating human intelligence in a machine, Wittgenstein glared up at the Darwin plaque over the mantelpiece, seeming to challenge the ghost of Charles himself to refute his argument in favor of the intrinsic uniqueness of man in his never-ending conflict with the machine.

Taking a long sip of brandy and a puff from his cigar, Snow stepped in to fill the painful silence that had descended upon the room following Wittgenstein's *cri de cœur*. Waving his cigar like a symphony conductor, he shifted the thrust of the discussion away from earthly matters of machines and humans by asking the group to direct their thoughts literally to the heavens.

"Perhaps we can gain some perspective on this man-versus-machine affair by considering how we might feel regarding things like human rights and the notion of personhood for an *alien* intelligence. Suppose, for the sake of discussion, that a spaceship from Andromeda landed tomorrow morning on Parliament Square and out jumped a completely alien being. To add a touch of piquancy to the situation, let me assume that this being is somehow able to speak to the Prime Minister in perfect English, but in physical appearance looks like one's worst nightmare—a hairy, mammoth-like creature having twelve tentacle-type legs with suckers at the end, a gaping hole for a mouth, antennae for ears and so forth. So what then? Would we be ready to confer personhood on such a human-like intelligence, even though its physical appearance differs radically from our own?"

Before discussion could even begin on this question, Wittgenstein loudly set aside his untouched glass of cognac and complained to Snow that he thought the question was meaningless. "The scenario you describe makes absolutely no sense at all, Snow. How can you imagine that these extraterrestrial beings could ever communicate in any fashion with us humans? Communication by language comes only from participation in a shared way of life. Look at the situation here on Earth. We cannot even communicate with creatures having a shared evolutionary history with us, such as monkeys, whales or termites. So how can you begin to think we might be able to communicate with beings that arose in a vastly different environment through a radically different evolutionary pathway? It's simply impossible, inconceivable even!"

"We have heard your social-group views on language and thought already, Wittgenstein," snapped Snow with a small show of irritation. "And they are surely worth serious consideration. But not all of us here share your opinion on this matter."

Raising his glass to attract Snow's attention, Turing hesitantly suggested that "Perhaps there is a kind of universal language that all technologically advanced intelligences must share. After all, Snow's hypothesis is that these Andromedans have built a spaceship in which they can travel between galaxies. This obviously requires a level of technological advancement that we can only dream about. I would say that there are certain facts about the physical universe that such technology would require these aliens to know—things like the value of π, the laws of celestial motion, the properties of atoms and so forth. It's not inconceivable to me that a language could be constructed around these universal facts that we could use to exchange *something* meaningful with these beings."

"Excellent point, Turing," added Schrödinger. "I'm sure the physics in Andromeda is the same as the physics here on Earth. And this fact may well serve as the basis for constructing a communication system that we both can share."

"Yes, I can't think of why the laws governing planetary motion or chemical reactions should be different in Andromeda," chimed in Snow. "So perhaps some sort of language based on counting or on atomic spectra might be workable as a means of cosmic communication. What do you think, Haldane?"

"All right, Snow, I concede that the laws of inanimate matter are likely to be no different in Andromeda than here on Earth. But that does not mean that the behavior of organisms like your mammoth-like, twelve-armed Andromedan is going to be anything at all similar to what we see here on Earth. So in this regard I'm sympathetic to Wittgenstein's position that what makes a human *human* is not our physical constitution. Rather, it is our social patterns. And these are extremely unlikely to be even comprehensible to an Andromedan, let alone be in any way similar."

"Well, how do you think they might differ from us?" asked Turing, curiously.

"My God," exclaimed Haldane, "there are so many ways they might differ that it's hard to list them all. Let me give you just a couple of examples. Suppose the Andromedans don't use money for the exchange of goods. Or suppose they ration children or practice incest? Such social patterns would be alien to us. But such behavioral patterns would form the basis for their way of life, hence their language, if we are to believe Wittgenstein. So how could we even begin to have a meaningful exchange of ideas with such creatures?"

Snow then asked: "Are you suggesting that if we cannot make cognitive contact with these Andromedans, it's then inconceivable that we could consider them to be 'persons' in the same sense as other humans?"

"Absolutely," asserted Haldane. "I think we may well regard them as creatures that deserve consideration, in much the same way we treat our cats and dogs with respect and affection. But we certainly would not think of these aliens as being 'persons'."

"Is there any difference between these beings from Andromeda and Turing's intelligent machines in this connection?" asked Schrödinger. "If the Andromedans are not persons, then I see no reason to accord that label to intelligent machines either," he concluded.

Taking another puff on his cigar, Snow looked over the top of his spectacles at Turing and asked: "Well, Turing, what do you say about this? Do you agree with Schrödinger and Haldane on this matter of personhood for your machines?"

Stammering slightly in his eagerness to express himself, Turing nervously replied: "I have never argued for or against this matter of personhood for machines. My concern has been solely with whether or not such a machine could be built that would display intelligence of the same order as that we see in our fellow human beings. Despite what I have heard here tonight, I still firmly believe that there is no logical barrier to the fabrication of such a machine. If you want to introduce moral and ethical issues, like personhood, machine rights and the like, I have no objection. But these other issues have nothing whatsoever to do with the question of machine intelligence."

"Well," said Snow, "it would appear we have just about come full circle on this business of the relationship between minds and machines. As the hour is getting late, perhaps I could ask each of you to briefly summarize your position on this matter. And just to be sure we are talking about the same question, let me state it again. Basically, it is what Turing just said: Is there any logical reason why we cannot envision technology advancing to the point where we could construct a computing machine that would be indistinguishable from a human being in its cognitive capabilities? Since you, Wittgenstein, have been the most vocal in your opposition to this question, perhaps I could ask you to begin by summarizing why you are against the very notion of a thinking machine."

Getting up from his seat on the sofa, Wittgenstein began to pace around the room, gathering his forces for a full-scale assault on Turing, Snow and anyone else who could even entertain this question as a matter of serious intellectual debate. Finally he turned to face the others, and spoke, softly at first, but with mounting intensity as his reply gathered momentum.

"The very idea of a machine thinking like a human is a total absurdity. It may well be possible to construct a machine that does well in Turing's Imitation Game, even fooling us into believing that it is thinking like you and me. But don't be fooled; as far as human thought goes, this would be a fraud. Human thinking is completely tied up with language, which in turn is a direct consequence of a shared form of life—human life. And no machine, regardless of how cleverly it is constructed, will ever be able to share that form of life just because it *is* a machine. So no, Snow, I do not believe at all in the possibility of a thinking machine."

"On philosophical, not technical grounds, right?" asked Snow, simply as a point of clarification.

"On any grounds the very question itself is meaningless," retorted Wittgenstein.

"All right," sighed Snow. "What do you say about the matter, Schrödinger?"

"I feel rather ambivalent about the whole business, Snow. On the one hand, I can't see any absolute logical or technical barrier to creating the kind of machine that Turing envisions. But I wonder what the point of such a device might be—other than to demonstrate our technical and engineering virtuosity? What is there to be gained by building such a machine?"

"Well, that's another issue altogether, and one that is perhaps best considered at another time," said Snow. "Right now, my concern is more with the feasibility of the idea as opposed to its actual worth."

Schrödinger then continued: "Yes, of course. In that case, I suppose I would have to side with Turing and say that I see no physical or technical reason why a machine could not be built that would fool us into believing it is thinking like a human."

Nodding his benevolent approval at this reply, Snow then asked Haldane to give his views on the possibility of a thinking machine.

"I'm afraid that I have to take a completely agnostic position on the issue. I just don't feel entirely comfortable with the idea of conferring such a basic human attribute as cognitive awareness on a mechanical gadget, no matter how cleverly constructed. I do feel that it's probably possible to build a machine that can do parlor tricks, and perhaps even fool us into accepting it as a thinking entity. But does this mean it is thinking *like* a human? I don't know. But I'd be extremely skeptical. There is something very special about the animal brain—and the human brain, in particular—and I'm very doubtful about being able to duplicate that specialness in a mechanical device. So when it comes to thinking like a human being, I would lean towards the ancient Scottish verdict of 'not proven'."

Noting Turing's reaction to these summary views, Snow asked him if he had any final thoughts on the matter that he would care to share with the group.

"Thank you, Snow. I don't have much to add to what I've already said on the matter during the course of this evening's discussion. But I do want to make clear the point that I have never felt that one of my machines would actually *duplicate* a human brain in a different medium. Some of you seem to have mistakenly interpreted my argument supporting a thinking machine as being tantamount to the creation of a copy of a human brain in metal, glass and plastic. That is not my goal at all. My beliefs—and goals—are far more modest. Put simply, my interest is in duplicating human thought processes, not human physiology. Now I accept the possibility that in order to do this it may be necessary to duplicate the human brain. And, of course, if this duplication were possible then presumably such a mechanical brain would by that very fact think just like a human. But my argument is that modern technology will enable us to capture human thought processes in a machine with much less than the complete duplication of a brain."

At that Snow put down his glass, stood up and addressed the room. "Gentlemen, the hour is getting late, and I think we have come about as far as we can for one evening on this question of thinking machines. Perhaps we can digest what's been said here tonight and meet again in a month or two to continue the discussion if our various schedules permit. But for now I think we have reached the time for reflection, not more debate."

"Hear, hear," added Haldane. "If that clock on the mantel is even approximately correct, I'm going to have to dash if I'm to make the last train back to Liverpool Street tonight."

As the group rose from their seats and moved to the hallway to collect their hats and coats, Snow said, "As I understood you earlier, Turing, you have arranged accommodation for the evening in your old rooms at King's. Is that right?"

"Yes, I confirmed it with the college porter on my way here earlier this evening," Turing replied.

"And you are staying with the von Wrights, Wittgenstein?" Snow enquired.

"Yes," Wittgenstein answered.

"I've asked the porter here at Christ's to make up the college guest room for you, Schrödinger. Will that suit you?"

"Very good, Snow," the Austrian replied. "I'm afraid it would be a bit of bother for me to go back down to London at the moment. So I'm afraid I'll have to let Haldane travel that route by himself this evening."

"Fine. So you are all taken care of for tonight," Snow said, ushering everyone into the vestibule. "Let me again thank each of you for taking the time to come up here to Cambridge this evening. Your views on the thinking machine question will certainly feature heavily in my final report to the Minister on the matter. I'm extremely grateful for your time. It has indeed been a splendid and intellectually rewarding evening for me. I hope you have all benefited from the discussion as much as I have."

"Indeed, I'm sure we have," replied Schrödinger. "I'm sure I speak for all of us when I say that I've seldom had such a fine dinner with such stimulating companions. Thank you very much for inviting us to this intellectual and culinary repast."

"My pleasure," responded Snow. "Have a speedy journey to your respective homes. And, again, thank you all very much for coming."

Closing the door to the clatter of his guests' footsteps on the stairwell as they made their way down to the courtyard, Snow leaned back against the door for a moment and breathed a sigh of both enormous relief and personal satisfaction. Making his way back to the sitting room, he helped himself to a generous refill of cognac and flopped down on to the sofa. What an evening, he thought. But how was he ever going to prepare a report to the Minister that would really do justice to the spectrum of ideas circulating in the room tonight?

How can I possibly compare Wittgenstein's philosophical position against thinking machines with Turing's technical arguments? And where do Schrödinger's and Haldane's views on the origin of life and the way living organisms behave fit into this overall picture? As if this were not enough, Snow mused, there's also Wittgenstein's Hieroglyphic Room thought experiment. As he laid it out, it certainly appeared to be a convincing refutation of the notion that a machine could ever think like a man. Somehow, though, the force of that argument was weakened by the discussion of thought and language that followed. How can I begin to convey the subtleties of these conflicting themes and views to someone like the Minister, who takes his science in small bites and with no rough edges? Well, he concluded, polishing off the last of the cognac, I've written reports before with a lot less information to draw upon than this. He got up from the sofa, stretched leisurely and then put out the lights and headed for the bedroom. The thinking-machine conundrum isn't going to be resolved in one evening, he thought, and tomorrow is another day . . .

Afterwards, ...

Afterwards, ...

C.P. Snow went on to achieve a certain level of intellectual visibility in 1959, when he coined the phrase 'the two cultures' in his Rede Lectures, to describe the gulf in world views separating the sciences and the humanities. He became a peer of the realm in 1964 and died in 1980. Haldane emigrated to India in 1957, partly in disgust over the British government's handling of the Suez Crisis. He continued his work on genetics there until his death in 1964. Schrödinger's little book, *What is Life?*, served as the impetus for the burgeoning field of molecular biology—despite the fact that his view of the structure of the gene was proved wrong by the work of Watson, Crick and others. Schrödinger finally returned to Vienna, where he lived the remainder of his life, dying in 1961.

Alan Turing died by his own hand in 1954, just three years after Wittgenstein succumbed to the prostate cancer that he was already suffering from at Snow's dinner party. So neither lived to see the dawning of the field christened 'artificial intelligence (AI)' by John McCarthy at the now-famous Dartmouth Conference in the summer of 1956.

By popular accord, the intellectual agenda for the global AI movement, even as it exists today, was set at this meeting at Dartmouth College. In attendance were such luminaries as Claude Shannon of information theory fame, Marvin Minsky, doyen of the MIT AI Lab, Frank Rosenblatt, pioneer of neural networks, Herbert Simon, Nobel-winning economist, as well as Simon's long-time Carnegie-Mellon University collaborator, Alan Newell. These people were to form the backbone of the AI research community as it spread across North America and the world in the 1960s.

Two basic approaches to the problem of machine intelligence were on the table at Dartmouth. The first, championed by Newell and Simon, took the view that cognition was a high-level phenomenon that could somehow be 'skimmed off' the brain, in much the same way that one skims cream off the top of a bottle of raw milk. The credo of this community is that intelligence is symbol processing in the brain. So, to create a similar intelligence in the machine requires simply creating the right silicon surrogates for the symbols the brain uses, and then generating the same rules the brain employs to shove these symbols around in the cranium. This is the essence of what came to be termed, 'Top-Down' AI. Forget about the actual physical structure of the brain, and focus on symbols and rules for their combination into new, and greater, symbols.

The loyal opposition at Dartmouth to the Top-Down approach was led by Frank Rosenblatt at Dartmouth, who emphasized the actual neuronal structure of the brain. This 'Bottom-Up' view said essentially that the actual structure of the human brain is important for carrying out its cognitive function, and therefore one should try to mimic this structure in hardware if you want to actually build a mechanical intelligence.

These two schools of thought on AI vied for supremacy during the early 1960s, until a freak event gave the upper hand to the Top-Downers. This event was the publication by Minsky and his colleague, Seymour Papert, of a result showing the impossibility of using one of Rosenblatt's neural models, the Perceptron, to solve a simple problem in boolean logic. For some unaccountable reason, this purely mathematical result was taken to mean that one could never emulate the activities of the human brain by a neural net, since the brain could easily solve this logical problem while the Rosenblatt machine could not. Following publication of this 'devastating' result, research funding for Bottom-Up AI dried up and students abandoned the area to work on the Top-Down approaches. A good chronicle of all this early history of the AI movement, Top-Down, and Bottom-Up, as well as an entertaining account of the various personalities involved can be found in the very illuminating and authoritative volumes, *Machines Who Think* by Pamela McCorduck (Freeman, San Francisco, 1979) and *The Mind's New Science* by Howard Gardner (Basic Books, New York, 1985).

During the 1960s and 1970s, research in AI centered on exploration of the Top-Down game plan. This involved a lot of different approaches to the problem of how to identify the right kinds of symbols of thought and rules for manipulating them, in order to get a computer to behave cognitively just like you and me. The biggest stumbling block in all these efforts was the so-called 'background intelligence' problem. As a result of human cognitive development from infancy, we all carry an enormous amount of background information around with us about how the world is; computers have no such catalogue of knowledge to draw upon, which makes it very difficult for a machine to understand a phrase like, 'The ball is in the pen.' Any human would immediately recognize the word 'pen' in this context to mean a child's playpen. But how to give this knowledge to a machine? Top-Down AI researchers still grapple with this problem today.

As a result of the meager progress made by Top-Down AI advocates on creation of a bona fide machine intelligence, together with the phenomenal advances in computer technology during the past twenty years, in the 1980s people began revisiting Rosenblatt's original Bottom-Up approach to AI via neural networks. But on the principle that if you revive an old idea you must give it a new name, workers labeled this line of investigation 'connectionism.' A rose by any other name . . . . An irreplaceable source for the philosophy of connectionism, as well as much more on minds, brains and machines is Douglas Hofstadter's Pulitzer-prize winning volume *Gödel, Escher, Bach: An Eternal Golden Braid* (Basic Books, New York, 1979).

In any case, it soon became clear that Minsky and Papert's result that killed work on neural networks in the 1960s was really irrelevant to the utility of the idea for thinking machines, and this fact, coupled with the widespread availability of cheap, powerful computing hardware led to a revival of research on Bottom-Up AI that continues to this day. A first-rate account of all this work on both Top-Down and Bottom-Up AI can be found in the easy-to-read volume, *Artificial Intelligence: A Philosophical Introduction* by Jack Copeland (Blackwell, Oxford, 1993).

The 1980s also saw the launching of two well-publicized broadsides against the very idea of a thinking machine. The first was philosopher John Searle's infamous Chinese Room argument, which is mimicked in the text by Wittgenstein's thought experiment involving the Hieroglyphic Room. Searle's arguments against the idea of the Turing test as a valid way of characterizing intelligence are vigorously presented in his book *Minds, Brains, and Science* (Harvard University Press, Cambridge, MA, 1984). The second major assault on strong AI was Roger Penrose's appeal to Gödel's Theorem in his best-selling book, *The Emperor's New Mind* (Oxford University Press, Oxford, 1989). I think it's safe to say that not many philosophers or scientists embrace either of these arguments, for all the reasons put forth by the guests at Snow's dinner party. So I won't re-elaborate them here. In retrospect, the real service done to the strong-AI community by both Searle and Penrose has been to galvanize the community into seriously re-thinking the philosophical issues underpinning their quest, and formulating effective counter-arguments.

As a final item in this fast-forward summary of AI research since Snow's time, let me say a word or two about the two touchstone problems that were held up as almost sacred missions by the founders of AI back at the time of the Dartmouth meeting nearly 50 years ago. These are the problems of computer chess-playing and natural language translation. Where does AI stand today on these fascinating questions?

In 1997, world chess champion Garry Kasparov was defeated in a tournament by Deep Blue–II, the reigning computer chess program. And, in fact, even on a home computer the good chess-playing programs now play at a level that only a human chess expert can expect to beat. So the claim made in the '50s that by the turn of the century the world chess champion would be a computer has now finally been realized. The real joke, however, is that in those days it was felt that producing a championship chess-playing program would somehow shed light on the way humans solve problems. Well, no such luck. In fact, what has been learned is that the way human grandmasters play chess and the way good programs play chess have almost nothing to do with each other. So the operation was a success—but the patient died!

What one can say, however, is that Deep Blue-II has passed a kind of Turing test for chess. And it's of considerable interest to note Kasparov's remark that he could see an 'alien intelligence' in Deep Blue's play. This is very different from the perspective of the program's designers, who know the machinery intimately, but cannot appreciate the subtlety of its play. So to Kasparov, who is able to appreciate its strengths, the program has become a kind of person.

All this notwithstanding, we have learned almost nothing about human cognitive capabilities and methods from the construction of chess-playing programs. A semi-technical, but still very readable, summary of this grand—but failed—experiment is provided in the volume *Kasparov versus Deep Blue: Computer Chess Comes of Age* by Monty Newborn (Springer-Verlag, New York, 1997).

In terms of everyday real-world competence, language-translation programs are vastly less successful than their chess-playing counterparts. But competence isn't everything. And when it comes to machines performing this most human of functions, as humans do it, machine translation projects have shown steady—if not remarkable—improvement in the past half century. Some of this progress has been due to a systematic investigation of the linguistic theories of Noam Chomsky that were introduced in the text. A good introductory source for the strengths and weaknesses of this theory is the volume *The Linguistic Wars* by Randy Harris (Oxford University Press, New York, 1993). Moving from painfully crude dictionary lookups to today's programs that do a passable job of producing rough, but usable, translations, there is some reason to hope, if not expect, that the day is not too far off when something akin to a Star-Trek-like universal translator will be realized. For some of the reasons why, the volume *An Introduction to Machine Translation* by W. John Hutchins and Harold L. Somers (Academic Press, London, 1992) is a good place to start.

If the half century of work on artificial intelligence has shown anything, it is that capturing human cognition within a machine is a very problematical affair. The things humans do well—pattern recognition, vision, inductive inference, creativity—machines do poorly, and vice-versa. This is not to say that human cognitive processes cannot be duplicated in a machine, but rather that it is a much trickier business than anyone thought in the 1950s. And many now feel that getting machines to think like humans is an exercise akin to getting robots to play football. It might be possible. But what's the point? It's like getting a horse to dance. Much more profitable is to recognize that this half a century of research has shown that these are two distinct forms of intelligence, and that for a short period they will peacefully coexist. After the current, but brief, interregnum, machines and humans will go their separate ways, much as humans and dolphins parted company many millennia ago. If he were alive today, I suspect that Turing would take a perverse satisfaction in seeing such a realization of his dream.

Photo Credits

Sources for the photos in the Dramatis Personæ section are:

Wittgenstein: Cover photo on the book *Recollections of Wittgenstein,* R. Rhees, ed., Oxford University Press, New York, 1984 (photo credit: Ben Richards)

Haldane: Cover photo on the book *On Being the Right Size and Other Essays,* J. Maynard Smith, ed., Oxford University Press, Oxford, 1985 (photo credit: Syndication International)

Turing: Last photographic plate in the book *Alan Turing: The Enigma* by Andrew Hodges, Burnett Books, London, 1983 (photo credits: the Royal Society and King's College, Cambridge)

Snow: Cover photo on the book *Stranger and Brother: A Portrait of C. P. Snow* by Philip Snow, Scribners, New York, 1982 (photo credit: Bern Schwartz)

Schrödinger: Cover photo on the book *Schrödinger: Life and Thought* by Walter Moore, Cambridge University Press, Cambridge, 1989 (No credit listed for the photo)